T0190091

Progress in Optical Science and Photonics

Volume 17

The purpose of the series Progress in Optical Science and Photonics is to provide a forum to disseminate the latest research findings in various areas of Optics and its applications. The intended audience are physicists, electrical and electronic engineers, applied mathematicians, biomedical engineers, and advanced graduate students.

More information about this series at https://link.springer.com/bookseries/10091

Xin He · Paul Beckett · Ranjith R Unnithan

Multispectral Image Sensors Using Metasurfaces

 Springer

Xin He
University of Melbourne
Melbourne, VIC, Australia

Paul Beckett
RMIT University
Melbourne, VIC, Australia

Ranjith R Unnithan
University of Melbourne
Melbourne, VIC, Australia

ISSN 2363-5096   ISSN 2363-510X (electronic)
Progress in Optical Science and Photonics
ISBN 978-981-16-7517-1   ISBN 978-981-16-7515-7 (eBook)
https://doi.org/10.1007/978-981-16-7515-7

This Springer imprint is published by the registered company Springer Nature Singapore Pte Ltd.
The registered company address is: 152 Beach Road, #21-01/04 Gateway East, Singapore 189721,
Singapore

To: Master James Ranjith, a curious mind.

Preface

The history of human civilization has been defined by our ability to manipulate and craft materials at ever decreasing geometries. We now find ourselves in the position of being able to manipulate materials at an atomic scale, even moving individual atoms around at will. This ability to arrange multiple materials at the nanoscale is opening up the possibility of completely new properties and functionalities in the way that materials interact with the physical world. The field of *Metamaterials* offers one such opportunity.

Metamaterials derive their basic properties from their inherent structure—their shape and size—such that their specific chemical makeup is of secondary importance. Although we often find that nature has beaten us to it, by manipulating these structures at nanoscale dimensions, novel functionalities may emerge that are not seen in natural materials. For example, it has been demonstrated many times that the refractive index of metamaterials can be "tuned" over a wide range from positive to negative values. Further, these effects are often achievable using thin-film 2D structures—*metasurfaces*—rather than necessarily requiring the use of volumetric or 3D metamaterials.

It is already clear that the advent of metamaterials is having a profound effect on the broad field of optics and optical sensors. The complex interactions between optical signals and metal–dielectric interfaces at nanometer scales—*plasmonics* or *nanoplasmonics*—support strong light–matter interactions and offer a promising way to develop simple yet effective narrow-band optical filters.

Since the first color digital camera came on to the market in the twentieth century, image sensor research has enjoyed a boom time. The image sensor is an essential part of any camera and serves to convert the optical signal to digital. This digital signal can be then further processed for color reconstruction, resulting in a color image. Multispectral cameras extend the concept of conventional color cameras to capture images with multiple color bands and with narrow spectral passbands. Images from a multispectral camera can extract significant amount of additional information way beyond the capability of the human eye or even a normal camera. Thus, they have important applications across a wide range of domains such as precision agriculture, forestry, medicine, and object identification and classification. Metasurfaces are opening up a range of new design possibilities in the domain of multispectral imaging systems. Even though establishing periodic arrays of nanostructures is difficult in itself, the technique does avoid the need for complex multilayer alignment that characterizes, for example, dye-based approaches.

On the other hand, it has so far been difficult to find materials and topologies that work across the visible and near-infrared (NIR) regions. Thus, while it is clear that metasurfaces offer a promising way forward, there is much work to be done before their full potential is realized. This book describes and analyzes a number of techniques for exploiting optical metasurfaces to create a single sensor-based multispectral camera architecture that works across the visible spectrum and into the NIR.

Melbourne, Australia   Xin He

August 2021   Paul Beckett

Ranjith R Unnithan

Acknowledgements

The work described in this book was supported in part by a University of Melbourne Interdisciplinary Seed Grant, the University of Melbourne McKenzie Fellowship Scheme, and through Australia Research Council Grant DP110100221. RRU also acknowledges additional ARC support through discovery project DP170100363.

The work was performed variously at the Melbourne Centre for Nanofabrication (MCN), the RMIT Micro-Nano Research Facility (MNRF) in the Victorian Node of the Australian National Fabrication Facility (ANFF), at the Materials Characterization and Fabrication Platform (MCFP) at the University of Melbourne, and the Victorian Node of the Australian National Fabrication Facility (ANFF).

The authors gratefully acknowledge the resources and services supplied by the National Computational Infrastructure (NCI), which is supported by the Australian Government.

We also acknowledge all of the contributors/authors of the journal papers used in the book and, in particular, assistance from Mr. Bryce Widdicombe during the drone tests.

Contents

Acronyms

AFM   Atomic Force Microscopy: a type of scanning probe microscopy (SPM) in which information is gathered by gently probing the surface with a nanoscale mechanical point.

ALD   Atomic Layer Deposition: a chemical gas phase thin-film deposition technique based on sequential surface reaction.

CFA   Color Filter Array: a mosaic of tiny color filters aligned over the pixels of an image sensor to capture color information.

CMOS   Complementary Metal Oxide Semiconductor: a type of field-effect transistor fabrication process that uses complementary and symmetrical pairs of p-type and n-type transistors

CMY   Cyan, Magenta, Yellow: a subtractive color model in which these three primary pigments or dyes are overlaid in various ways to reproduce a range of colors.

DCHA   Dual Coaxial Hole Array: plasmonic structures with an array of subwavelength coaxial apertures in a metal film.

DRIE   Deep Reactive-Ion Etching: a highly anisotropic etch process used to create deep, steep-sided holes, and trenches in wafers and/or substrates.

EBL   Electron Beam Lithography: manufacturing process involving the scanning of a focused beam of electrons to draw custom shapes on a surface covered with an electron-sensitive film.

EM   Electromagnetic: refers to an electromagnetic field propagating through space carrying electromagnetic radiant energy.

FEM   Finite Element Methods: a method for numerically solving differential equations.

FFOV   Full Field of View: alternative to FOV used to emphasize that it is a diameter measure.

FIB   Focused ion beam: a technique employing a focused beam of electrons for the analysis, deposition, and ablation of materials.

FOV   Field of View: the solid angle through which a detector is sensitive to electromagnetic radiation.

FSS   Frequency-Selective Surfaces: any regularly patterned surface designed to reflect, transmit, or absorb electromagnetic fields at selected frequencies.

FWHM   Full Width at Half Maximum: width of a spectrum measured between its 50% maximum amplitude points.

GaAs   Gallium Arsenide:a III-V direct band gap semiconductor with a cubic crystal structure used for high-frequency integrated circuits and optical devices.

Gamut   Color Gamut: defines a specific range of colors from the range of colors identifiable by the human eye (i.e., the visible spectrum) and is intended to clarify and quantify the common variations in color produced by different image capture devices.

GMRF   Dielectric Guided Mode Resonance Filters: frequency selective system that uses a phase-matching element, such as a diffraction grating or prism, to excite and simultaneously extract the guided modes within an optical waveguide.

LCP   Left-hand Circular Polarized: one of two circular polarization states defined by the direction in which the electric field vector rotates relative to the specific point of view adopted (cf. right-handed, RCP).

LSP   Localized Surface Plasmon: the result of the confinement of a surface plasmon in a nanoparticle of size comparable to or smaller than the wavelength of light used to excite the plasmon.

MIBK   Methyl Isobutyl Ketone: a liquid organic hydrocarbon commonly used as a solvent for gums, resins, paints, varnishes, lacquers, and nitrocellulose.

MIS   Multispectral Imaging System: a complete system of lenses, filters, sensors, etc., that captures image data within specific (often narrow) wavelengths across the electromagnetic spectrum.

MS   Multispectral Camera: the image acquisition component of a MIS.

NASA   National Aeronautics and Space Administration.

NIR   Near Infrared: defined approximately as a region of wavelengths between 750 and 2500 nm.

PBC   Periodic Boundary Condition: a set of boundary conditions used in computer simulations and mathematical models which are often chosen for approximating a large (infinite) system by using a small part called a unit cell.

PC   Photonic Crystals: periodic optical nanostructures that are designed to form the energy band structure for photons, either allowing or preventing the propagation of electromagnetic waves within specific frequency ranges.

PD   Photodetector: also photodiode (used interchangeably), the photon sensor for both conventional digital cameras and MIS.

PEC   Perfect Electric Conductor: an idealized material exhibiting infinite electrical conductivity or, equivalently, zero resistivity.

PECVD   Plasma-enhanced Chemical Vapor Deposition: a chemical vapor deposition process used to deposit thin films from a gas state (vapor) to a solid state on a substrate.

PMC   Perfect Magnetic Conductor: a hypothetical material containing no internal magnetic field components used to find approximate solutions for some problems in electromagnetics.

PML   Perfectly Matched Layer: artificial material used to create totally absorbing boundary conditions, applied to FEM simulations with open boundaries.

PMMA   Poly (Methyl Methacrylate): also called acrylic, acrylic glass.

RBVI   Red–Blue Vegetation Index: standardized mathematical equations applied to the spectra of light reflected from vegetation, designed to maximize sensitivity to the vegetation characteristics while minimizing confounding factors such as soil background reflectance, directional, or atmospheric effects.

RCP   Right-hand Circular Polarized: see LCP.

RGB   Red, Green, Blue: additive color space similar to the function of the eye.

SBC   Scattering Boundary Condition: boundary conditions applied when solving wave electromagnetics problems that are transparent to all outgoing waves in the limit that it is infinitely far away.

SOG   Spin on Glass: a mixture of $SiO2$ and dopants (typically boron or phosphorous), suspended in a solvent solution and applied to a surface by spin coating.

SPP   Surface Plasmon Polaritons: electromagnetic waves that travel along a metal–dielectric or metal–air interface and so involves both charge motion in the metal (i.e., surface plasmon) and electromagnetic waves in the air or dielectric (i.e., polariton).

SSMC   Single Sensor-based Multispectral Camera: a MIS system in which a single image sensor is used to acquire multiple spectral components.

SWIR   Short Wavelength Infrared: typically defined as light in the 900–1700 nm wavelength range, but can sometimes be used to refer to the same 700–2500 nm region as NIR.

TE   Transverse Electric Mode: electromagnetic radiation is in the plane perpendicular to the direction of propagation.

TM   Transverse Magnetic Mode: magnetic field of a wave is entirely perpendicular to its direction of propagation.

t-SPL   Thermal Scanning Probe Lithography: a form of scanning probe lithography (SPL) where the material is structured at nanoscale dimensions using scanning probes, primarily through the application of heat.

UAV   Unmanned Aerial Vehicle (also known as a drone): any aircraft with no human crew or passengers, able to operate under various degrees of autonomy from remote operator control, partial control with autopilot assistance to be fully autonomous.

Chapter 1
Introduction to Metasurfaces for Optical Applications

For most of human history, we have been constrained to manipulate and shape materials in a way that imitates nature. This has arguably been an extremely successful strategy and our advanced civilization is built on this ability to harness natural phenomena for our own purposes. Around the year 2000, it became possible to engineer materials at scales less than 100 nm and we entered what has become known as the 'nano-' age: so we now speak of nano-materials, nano-electronics, nano-photonics etc. The ability to arrange multiple materials at the nano-scale has opened up numerous possibilities to create systems with completely new properties, and/or functionalities not previously seen in nature. One such opportunity is the topic of this book: *metamaterials* or, more specifically, their 2-D counterparts *metasurfaces*.

In general terms, metamaterials are synthetic materials in which their geometry (shape, size, orientation, separation etc.) is the primary determinant of their electromagnetic properties. As such, they are not constrained in the same way as conventional matter and can exhibit properties that do not occur naturally. Commonly cited examples include materials that show a negative index of refraction. The key idea here is that, although metamaterials are constructed from conventional metal and dielectric materials, it is the topological arrangement of these into repeating ('unit') cells that induces localized interaction between the cells and an incident electromagnetic field, thereby determining the material's macroscopic behavior. In turn, the cell dimensions determine the frequency of operation of the material with the unit cell size being typically much shorter than the operating wavelength. Thus, metamaterials with unit cells in the range of a few millimetres will operate at microwave frequencies whereas to work within the visible part of the spectrum requires typical dimensions of a few nanometres. For this reason, while the theoretical properties of metamaterials were first described more than 50 years ago, it has only become possible to reliably manufacture photonic metamaterials in the last twenty years or so.

The concept of a *metasurface* [1] extends this idea further with the observation that many of the characteristics of metamaterials, particularly in the optical domain, can be achieved using a 2-D nanostructured interface with subwavelength thickness. Although the idea of a metasurface is relatively recent, similarly engineered surfaces have already been in wide use across a range of applications such as frequency-selective surfaces (FSS) in the microwave region and gratings and photonic crystals in the optical domain [2]. Metasurface-based techniques also avoid some of the drawbacks of 3-D metamaterials such as high loss. Manufacturing may also be easier as many of the techniques for building nano-scale surfaces have been in common use for many years in the micro-electronics industry.

It is already evident that the application of metasurfaces is having a great impact right across the field of optics and imaging. This is particularly true for multispectral imaging, which relies on the ability to produce optical filters exhibiting extremely sharp passbands, something that is difficult to achieve using conventional optical filtering techniques.

This chapter sets up the theoretical foundations for the rest of the book, which will examine the application of metasurface techniques to narrow-band optical filtering. It begins by looking at the theory behind conventional lenses in which their overall behavior is determined by accumulated phase and amplitude changes that occur as light propagates through the material. Such lenses are therefore constrained by the refractive index of their materials and their bulk shape (concave, convex). This contrasts with the second theme of this chapter, metasurfaces, where the macroscopic behavior emerges from surface features with sub-wavelength dimensions and/or the abrupt boundaries between materials of different refractive indices. The chapter therefore covers the basic theory used to describe and model both conventional and metasurface-based optical systems that will be used later in the book.

1.1   Conventional Optical Models

As mentioned above, conventional lenses rely on the large-scale geometry and refractive index of the material making up the lens. This behavior is illustrated in Fig. 1.1a and can be modeled using the classical set of Eq. (1.1) known as Snell's Law:

$$\theta_1 = \theta_3$$
$$n_1 \sin \theta_1 = n_2 \sin \theta_2. \tag{1.1}$$

If $n_1 > n_2$, when θ_1 reaches the point where the angle of refraction is 0, total internal reflection occurs and no energy is coupled into the second material (Fig. 1.1b).

As an example of this, Fig. 1.2 describes a scenario where monochrome light with a wavelength λ passes through an optical system based on three dielectrics that exhibit two different values of refractive index (n_1 and n_2, with $n_1 > n_2$). At each boundary a fraction of the energy will be reflected and, ignoring losses, the remainder propagated into and through the dielectric medium (note: for clarity, not

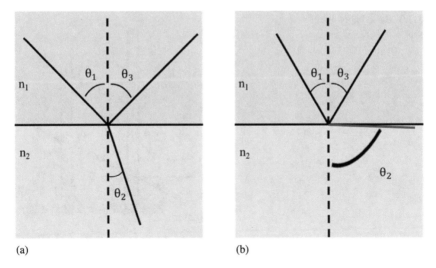

(a)   (b)

Fig. 1.1 Behavior of light in dielectrics. **a** Reflection and refraction of light at the interface between two media. **b** Total international reflection and critical angle

Fig. 1.2 An example of an optical system with two different refractive indices

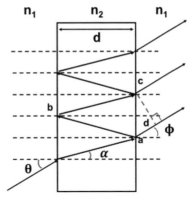

all rays are shown here). It can be seen that the distance a specific ray travels will depend the dielectric thickness d and the angle α, which determines the internal reflection path. Each of the rays leaving the center dielectric will have an additional length $md\cos(\alpha)$ ($m \in \{1, 2...\}$) added to its path. The optical path difference (Δ) is defined as the difference in the lengths that of two optical paths while passing through materials with different refractive indices and is therefore given by:

$$\Delta = n_2(ab + bc) - n_1(ad) = 2n_2d\cos(\alpha). \tag{1.2}$$

As a result, each exit ray is phase-shifted by an amount proportional to the optical path difference. If we consider two electromagnetic (EM) waves with the general expressions :

Fig. 1.3 Basic Fabry–Pérot
Etalon

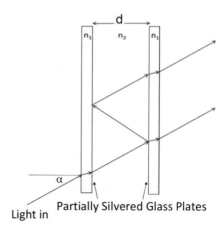

$$E_x = E_{x0} \sin(kx - \omega t + \eta_x)$$
$$E_y = E_{y0} \sin(ky - \omega t + \eta_y), \qquad (1.3)$$

where $\omega = 2\pi f$ is the angular frequency, $k = 2\pi/\lambda$ is the wavenumber, the phase shift η between the two is the sum of the initial phase difference plus that due to the optical path and:

$$\eta = \frac{2\pi}{\lambda}(x - y) + (\eta_x - \eta_y). \qquad (1.4)$$

For a system such as the example in Fig. 1.2, the phase shift due to the optical path difference, referenced to the phase of the initial ray can be written as:

$$\eta = \frac{2\pi}{\lambda}\Delta = \frac{4\pi n_2 d \cos(\alpha)}{\lambda}. \qquad (1.5)$$

This phase shift mechanism is widely exploited in optical systems and forms the basis for the operation of convex and concave lens. An equivalent phase shift can be generated using metasurfaces, as will be introduced in Sect. 1.2.

1.1.1   The Fabry–Pérot (FP) Etalon

The classic structure of the Fabry–Pérot (FP) etalon,[1] shown in Fig. 1.3 [3] comprises a single optical cavity, or dielectric plate, with two parallel (partially) reflecting surfaces. It can be seen that this is essentially identical to the multilayered organization described in Fig. 1.2 where the cavity, or center dielectric has a smaller refractive

---

[1] Although the terms are used interchangeably, strictly, an *etalon* refers to the case where the resonance spacing, d, is fixed. The device tends to be called an *interferometer* when d is variable.

index (n_2) than two outer layers (n_1). As before, at each boundary the energy of an incident ray splits into reflected, transmitted and loss components. The spectral response of a Fabry–Pérot resonator is based on interference between the incident light and that reflected (and therefore phase-shifted) at the mirror surfaces. The phase shift can be derived by considering the point of maximum power transmission $\eta = 2m\pi$, for integer m. As the path difference between two adjacent refracted light rays is defined as:

$$\Delta = n_2(ab + bc) - n_1(ad) = 2n_2d\cos(\alpha), \qquad (1.6)$$

and the relative phase change is identical to the previous case and is given by (1.5), above. Thus, the transmitted power will be maximized when $\nu = 2m\pi$, for integer m > 0. If it is further assumed that the angle of incidence is 0, then the value of α in (1.5) will be 0 so that the peak wavelength becomes:

$$\lambda = \frac{2n_2d}{m}, \qquad (1.7)$$

where, as before, m = 1,2,3... etc. The peak transmittance is therefore periodic in n_2d and can be relatively easily tuned by a choice of dielectric material and spacing. This concept forms the basis for much of the existing narrow-band optical filter work and will be revisited in Chap. 2.

1.1.2 Diffraction Gratings

Due to its inherent simplicity, the diffraction grating is probably the device most widely used to modify the characteristics of an incident electromagnetic wave. Put simply, reflection off or transmission through a regular grating structure results in constructive or destructive interference which, in turn, imposes on the incident wave a similarly regular (periodic) variation in amplitude and/or phase [4]. For a more complete analysis of diffraction grating behavior, the reader is referred to [5]. Grating structures are very commonly used as dispersive elements across a range of applications in spectrography, as beam splitters and combiners in interferometers and laser-based optical systems, in optical modulators or even as non-reflective surfaces [6–10].

The general theory of diffraction gratings describes how any optical system based on a periodical structure can diffract light into multiple scattering orders (i.e., directions). A transmission grating will diffract monochromatic light with wavelength λ through an angle θ_m according to [5]:

$$n_2\sin\theta_i - n_1\sin\theta_m = m\lambda/d, \quad m \in \mathbb{Z}, \qquad (1.8)$$

where θ_i is the angle of incidence, n_1 and n_2 are the refractive indices of the two adjacent media along the path of the ray, and m is an integer known as the *order* of diffraction. The equation for the reflection case can be found simply by setting $n = n_2 = -n_1$ in (1.8) so that:

$$\sin \theta_i + n_1 \sin \theta_m = m\lambda/nd, \quad m \in \mathbb{Z}, \tag{1.9}$$

and the zeroth order (m = 0) corresponds to the directly transmitted or specularly reflected beam. Peaks in the diffracted wave occur at angles defined by the non-zero integers m. It can be seen from (1.8) and (1.9) that the grating period, d, is the key parameter that limits the dispersion and resolution of the system, which leads to a trade off between the period and spacing of a grating versus its transmission efficiency.

A number of important applications of diffraction gratings are based on their ability to provoke strong resonance at the boundary between dielectric surfaces. Conventional examples to date include planar waveguide couplers and optical multiplexers [11]. However, replacing the dielectric waveguide by a metal can induce *plasmonic* behavior in which surface plasmons are almost totally confined to the interface region between air and the metal surface [12] and are therefore guided along the interface layer. Considering only the first scattering order, the propagation constant of a metallic nano-grating is given by:

$$\beta = \pm\frac{2\pi}{P} + k_0 \sin \theta, \tag{1.10}$$

where P is the period of the grating, θ is the angle of incidence, and $k_0 = \frac{2\pi}{\lambda}$ is the wavenumber of the incident light in air. If the propagation constant of the surface plasmon matches that of the localized evanescent wave, an incident beam interacting with even a very thin metallic grating may generate surface plasmons by resonant excitation, resulting in large coupling efficiencies at very specific wavelengths and angles of incidence. This plasmonic resonance phenomenon is essentially the same mechanism that explains the strongly wavelength dependent transmission of light by sub-wavelength metallic hole arrays to be discussed in detail in Chap. 3.

1.2 Metasurfaces Applied to Conventional Optics

In the past decades, metasurfaces based on the sub-wavelengths micro-nanostructures have started to replace conventional methods of manipulating the optical properties of light, such as its amplitude, wavelengths and polarization status. As outlined in the introduction above, metasurfaces are formed from metallic or dielectric structures with feature sizes much smaller than the operational wavelengths. They can be used for a broad range of optical applications in a similar manner to conventional optical devices, and offer a number of unique advantages such as portability and flexible customization. They can be more robust than organic based devices in harsh environments. For example, while most polymer based optical devices are easily damaged by laser energy, metal and dielectric thin films may be more resistant.

In this section, we examine a small number of the ways that metasurfaces have been applied to standard optics. Firstly, TiO_2 nano-fins with thicknesses in the range of hundreds of nanometers exhibit sufficient path differences to be formed into so-called *meta-lens* structures that exhibit equivalent characteristics to conventional lenses. Further, chiral metasurfaces can replace the relatively complex optical setups required to image polarized light, in particular greatly simplifying the separation and acquisition of the circularly polarised components of the incident light [13]. Metasurfaces with sub-wavelength features have also been shown to filter visible spectra, in this case using the interaction between the incident light and surface plasmons. These plasmonic optical filters will be introduced at the end of this section.

1.2.1 Meta-Lenses

Conventional optical lenses use the macroscopic geometry of their dielectric material to change the optical path. These lenses tend to be bulky and highly curved and are subject to a number of problems such as distortion, flare, ghosting, aberrations, and so forth. In fact, while high quality camera lenses are deliberately made very large in order to overcome some of these problems, it is still often necessary to post-process the resulting digital images to remove distortion, edge effects and the like that arise due to the geometry of the lens.

The metasurface concepts outlined in the previous section can be applied to fabricate *meta-lenses* with more-or-less equivalent characteristics but that are geometrically flat. The basic principle is to mimic the refractive behavior in a conventional concave or convex lens by engineering appropriate optical phase changes across the surface of the meta-lens. Light rays passing through different points on the surface will 'see' different optical path lengths (and therefore undergo different phase shifts) so that the overall structure will behave like a lens.

Many proposals to effect the necessary phase changes have been presented, using various dielectric and/or metallic nanostructures [14–19]. For example, it has been shown in [14] (Figs. 1.4 and 1.5) that by rotating a thick TiO_2 nano-fin from 0 to 180°C, the phase can be shifted from 0 to 2π. In this case, the thickness of the TiO_2 nano-fin (Fig. 1.4b) needs to be similar to the operating wavelength (i.e., $\sim$400–700 nm to cover the visible spectrum). When the nano-fins are arranged with 0° rotation in the center, increasing to 10°, 20°, . . . through to 180° in successive outer rings, the overall structure behaves like a convex lens. Configured in the opposite arrangement (i.e., with 180° rotation at the center), the surface acts as concave lens. This meta-lens surface is flexibly scalable—it can be made as small as a hundred micrometers up to a range of multiple centimeters.

Metallic nano-structures based on plasmonic theory have also been shown to support sufficient phase shifts at optical frequencies to be usable as meta-lenses.

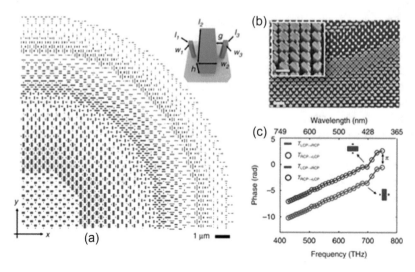

Fig. 1.4 Metalens based on TiO_2 nano-fin rotation. From [14], licensed under a creative commons attribution (CC BY 2.0).

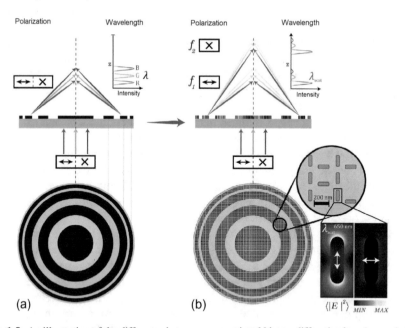

Fig. 1.5 An illustration of the difference between conventional binary-diffractive lens (zone plate) and the plasmonic enhanced lens described in [15] for orthogonal linearly polarized input. **a** Binary transmittance regions composed of a thin opaque material, such as chromium. **b** Plasmonic-enhanced flat-lens (metalens) composed of subwavelength scattering elements (metallic nanostructures). Reproduced from [15], licensed under a creative commons attribution (CC BY 2.0)

Their key advantage is that the thickness of a plasmonic meta-lens can be as low as 30 nm, much smaller than the corresponding dielectric structure. However, all of the dielectric materials proposed to date are inherently transparent, whereas the metallic materials exhibit higher values of reflection and absorption. Thus the transmission efficiency of plasmonic devices is much lower than that of dielectric-based systems.

Although flat meta-lenses are significantly smaller than conventional lenses and are inherently more flexible to design and build (avoiding, for example, the need to fabricate complex 3-D shapes in the lens material), they still have a number of critical disadvantages that limit their usefulness. Key amongst these is the issue of chromatic aberration. By their very nature, the dielectric or metallic structures forming a meta-lens will work optimally at a particular wavelength, or across a very narrow range, and their performance will degrade rapidly at other wavelengths, a characteristic that will prevent their use in applications that require operation across a broad optical spectrum.

1.2.2 Metasurfaces for Polarization Imaging

Polarization is an essential quality of electromagnetic radiation and describes the geometric orientation of the oscillation of its electric field. The coupled, perpendicular magnetic and electric fields of an optical wave are referred to as *linearly* polarized if they oscillate in a single direction and *circularly* (or, more generally, *eliptically*) if they rotate at a constant rate along the direction of travel. The optical path of circularly polarized light in 3-D is a helix oriented along the direction of propagation, a structure also referred to as *chiral* to reflect its intrinsic "handedness". Circular polarisation (CP) therefore exists in "left-hand" (LCP) and "right-hand" (RCP) forms, although the actual definition of the "hand" will depend on the adopted viewpoint, either sender or receiver.[2]

Conventional camera sensors measure intensity only and so optical polarization is invisible to them. On the other hand, as polarization can be influenced in deterministic ways by scattering within a medium, it can reveal important features not otherwise discernible in an image. For this reason, polarized imaging is widely used in biological, chemical, aerial imaging and other research fields.

A complete description of the polarization state of a light ray is given by its Stokes vector, which can describe a full range of un-polarized, partially and fully polarized light. The full Stokes vector requires at least four individual measurements, which in the past has meant that these optical systems have been bulky and often reliant on moving parts.

Starting with the general EM expressions in (1.3), above, the phase difference between these two waves can be written as:

---

[2] Both reference viewpoints are in common use in different contexts. A sender (source) viewpoint is typical in Engineering, Physics and Astronomy and is an IEEE standard, while the receiver viewpoint tends to be used within the optical and chemical domains.

Table 1.1 Table of primary polarizations

	LHP	LVP	LP+45	LP-45	LCP	RCP
E_{x0}	E_0	0	E_0	E_0	E_0	E_0
E_{y0}	0	E_0	E_0	E_0	E_0	E_0
η_Δ	0	0	0	π	$\frac{\pi}{2}$	$-\frac{\pi}{2}$

Fig. 1.6 Poincaré Sphere and six degenerate polarization states

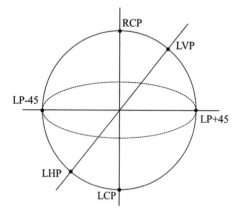

$$\frac{E_x^2}{E_{x0}^2} + \frac{E_y^2}{E_{y0}^2} - \frac{2E_x E_y}{E_{x0}E_{y0}}\cos(\eta_\Delta) = \sin^2(\eta_\Delta). \tag{1.11}$$

This allows us to define six primary directions of polarization. as shown in Fig. 1.6 and Table 1.1, which in turn represent the key components to calculate the Stokes parameters. These six points are called the *degenerate* polarization states and can be represented as a point on the surface of the Poincaré sphere as shown in Fig, 1.6.[3]

To represent full Stokes polarization images, the necessary four parameters, $S_{0...3}$, can be measured and calculated with [18]:

$$\begin{aligned}
S_0 &= I_0, \\
S_1 &= I_{0°} - I_{90°}, \\
S_2 &= I_{45°} - I_{-45°}, \\
S_3 &= I_{RCP} - I_{LCP}.
\end{aligned} \tag{1.12}$$

S_0 represents the signal intensity of un-polarized light and can be measured with a monochrome camera and lens. S_1 is the difference between two linear values at polarization angles of 0° and 90°. S_2 can be measured with a monochrome camera integrated with 45° and −45° linear polarizers. S_3 is the value between the right

---

[3] It is worth noting that both circular and linear polarization can be considered to be special cases of the more general elliptical form that traces out at an angle to the Cartesian coordinates in Fig. 1.6 However, we limit the discussion in this book to the linear and circular forms.

and left circular polarization intensities [18]. After measuring these four Stokes parameters, they can be mapped to the 3D Poincaré sphere, as shown previously in Fig. 1.6.

Linear polarizers can be designed using metallic nano-gratings. This technology is well-developed and many commercial products are already available (e.g. Thorlabs [20] and Edmund Optics [21]). In contrast, circular polarizers based on metasurfaces are still an active area of photonic research because conventional polarization imaging typically requires a complex optical path including items such as a phase retardation plate, linear polarizer and a monochrome camera, all of which have to be set up and carefully aligned. In contrast, metasurface techniques can be very compact and can remove the need for much of this hardware.

There have been a number of approaches already proposed to the development of metasurface-based circular polarizers. For example, as well as acting as a bandpass filter, the dielectric nano-fin array discussed in Sect. 1.2.1 above can also behave as a diffraction grating [19]. Note that in this case, the thickness of the dielectric surface is of the same order as the optical wavelengths ($\sim$400–700 nm).

Four polarization components of the incident light can be diffracted into four directions by metasurfaces with different polarization angles and phase differences. By collecting the signal intensity of each, the Stokes parameters of the polarized light can be measured and its status marked on the Poincaré sphere.

As mentioned above, the optical path of circularly polarized light is chiral so that the normalized Jones vectors for the two orthogonal CP states will be given by: $\overrightarrow{\lambda}_{RCP} = \frac{1}{\sqrt{2}}\binom{1}{i}$ and $\overrightarrow{\lambda}_{LCP} = \frac{1}{\sqrt{2}}\binom{1}{-i}$ for the right- and left-hand states, respectively. The objective of the chiral metasurface, therefore, is to shift these opposite states by known angular displacements from the optical axis so the polarization components can be either passed, blocked or separately acquired. In [22], a process of electro-chemical deposition was used to fabricate three-dimensional gold helices arranged on a two-dimensional square lattice. The resulting structures block circularly polarized light exhibiting the same handedness as the helices, while passing the other. The filter operates over a relatively wide frequency range, albeit in the middle to far—infrared regions between 2.5 and 5.6 μm, rather than in the visible. A compact meta-lens element was proposed and simulated in [23], using rectangular-shaped TiO_2 pillars. The proposal combines propagation and geometric phases to allow the LCP and RCP to be shifted independently of each other.

As these are predominately metal structures, chiral metasurfaces can also be analysed using plasmonic theory. Metallic gratings are polarization sensitive so that rotating them continuously in 3-D will result in a helix structure that will pass only LCP or RCP components [24].

However, this type of polarization imaging system has a number of drawbacks, including:

- as the operating wavelength is largely determined by the (fixed) geometry of the structure, the optical system typically works optimally at only one wavelength, or over a narrow spectral band, although a multilayer stack approach was presented in [25] which covers a wider wavelength range in the visible, from 300 nm.

- each of the components are diffracted at a different angle to the optical axis and therefore require additional optics to focus the diffracted beams, meaning that the optical measurement path might be as complex as a conventional setup;
- in general, the chiral structures are quite difficult to manufacture. For example, the multilayer stack in [25] needs multiple overlaid stages of lithography, resulting in a complicated manufacturing process.

1.2.3 Metasurface-Based Plasmonic Filters

As has been already identified, an important characteristic of metasurfaces i.e., geometric arrays of repeating sub-wavelength cells in a thin metallic layer over a dielectric substrate, is their ability to produce and tune various plasmonic resonance modes that depend on the size, shape and periodicity of the repeating cells. In particular, metasurfaces can be set up to exhibit and exploit localized surface plasmon resonance (LSPR) [26], surface plasmon polariton (SPP) [27], extraordinary transmission [24], surface lattice resonance (SLR) [28], Fano resonance [29], plasmonic whispering-gallery modes (WGMs) [30], and plasmonic gap mode [31]. As a result, the absorption, scattering, transmission, and guided wave propagation of light from and through a medium can be controlled by the deliberate design and fabrication of plasmonic array metasurfaces. For a more complete description of progress in surface plasmon resonances of 2-D and 3-D nanostructure array patterns, the reader is referred to [32].

As briefly introduced in Sect. 1.1.2, surface plasmon polaritons (SPP) occur if a plasmon is excited and propagated at the interface between a metal and dielectric. For a photon to excite a surface plasmon polariton, both must have the same frequency and momentum. Thus, the coupling of photons into SPPs can be achieved using any method that allows the momentum of one to be matched to that of the other. Typical mechanisms include prisms or gratings to match the photon and SPP wave vectors, thereby also matching the momentum of each. Additionally, localized surface plasmons (LSP) arise when the excited plasmons are not propagating at the surface of the metallic nanoparticle (e.g. nano-sphere, nano-disk) but are still coupling to the electromagnetic field [33].

For SPPs, the propagation constant can be written as:

$$\beta = k_0 \sqrt{\frac{\varepsilon_1 \varepsilon_2}{\varepsilon_1 + \varepsilon_2}}, k_0 = \frac{\omega}{c} = \frac{2\pi c/\lambda}{c} = \frac{2\pi}{\lambda}, \tag{1.13}$$

where ε_1 and ε_2 are the permittivity of the metal and dielectric respectively, ω is the angular frequency, λ the wavelength, and c is the speed of light in a vacuum. It can be seen that the propagation of SPPs is largely a function of the materials. In contrast, the peak wavelength for the LSP case mostly depends on specific dimensions of the surface features. Considering the propagation constant of the metallic grating structure given previously in (1.10), for the case where the light is normal to the sensor surface, $\theta = 0$, so that the wavelength (λ) is proportional to the grating period (P):

$$k_0 \sqrt{\frac{\varepsilon_1 \varepsilon_2}{\varepsilon_1 + \varepsilon_2}} = \frac{2\pi}{P}, \lambda = P \sqrt{\frac{\varepsilon_1 \varepsilon_2}{\varepsilon_1 + \varepsilon_2}}. \tag{1.14}$$

There are two arrangements in common use to create arrays of metallic nano-hole: square and hexagonal. The propagation constant (β) of both of these topologies can be described as [24]:

$$\beta = K_0 \sin \theta \pm n G_x + m G_y = K_0 \sin \theta \pm (n + m) \frac{2\pi}{P}, \tag{1.15}$$

where where $K_0 \sin \theta$ represents the component of the incident photon's wavevector in the plane of the surface, $G_x = G_y = 2\pi/P$ are the momentum wavevectors for the square array, with n and m the scattering orders. The peak wavelength (λ) for a square array is given by:

$$\lambda = \frac{P}{\sqrt{\frac{4}{3}(n^2 + m^2)}} \sqrt{\frac{\varepsilon_1 \varepsilon_2}{\varepsilon_1 + \varepsilon_2}}, \tag{1.16}$$

and for a hexagonal hole array by:

$$\lambda = \frac{P}{\sqrt{\frac{4}{3}(n^2 + nm + m^2)}} \sqrt{\frac{\varepsilon_1 \varepsilon_2}{\varepsilon_1 + \varepsilon_2}}, \tag{1.17}$$

with scattering orders n and m and the permittivity of the two materials, $\varepsilon_{1,2}$, as above.

An illustrative example of a square array organization from [34] is shown in Fig. 1.7 for which a number of transmission peaks are evident. The first at around 500 nm (labeled λ^* in Fig. 1.7) corresponds to the combination of interband

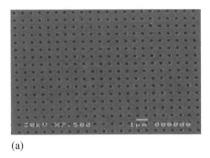

(a)

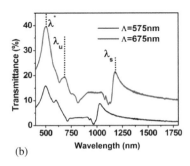

(b)

Fig. 1.7 **a** A SEM view of a square array of Au nano-holes fabricated by double exposure of two E-beam interference patterns and **b** the resulting transmission spectra (in air) of two samples with hole diameter/period of 230/575 and 250/675 nm. Reproduced from [34] under a creative commons attribution licence (CC BY 4.0)

absorption with the free carrier response and is therefore independent of the hole geometry. The remaining peaks are due to the two metal-dielectric interfaces: a minor peak from the metal-air boundary (λ_u) and a much stronger peak due to the metal-substrate (λ_s) interface. Both show a dependence on the period of the hole array[4] and the peak wavelengths can be predicted from (1.16). Although the 675 nm (red) curve in Fig. 1.7 is offset for clarity, so that the *absolute* value of transmittance for that curve is not relevant, it is still clear that the peak values tend to be very low (below 20% in all cases). This is the main objective of the technique presented in Chap. 3, in which uses a coaxial hole array is used to tune both the localized surface plasmons and surface plasmon polaritons to greatly enhance the transmission efficiency of the resulting plasmonic color filters. Further, in Chaps. 5 and 6, this concept is extended to produce a number of very narrow band color mosaics formed from a hybrid combination of plasmonic color filters and heterostructured dielectric multilayers. These hybrid filters are both readily tunable and can produce very narrow spectral spectral width (FWHM) values, typically smaller than 50 nm and down to as low as 17 nm at some wavelengths.

1.3   Summary

This chapter has presented a brief introduction to metasurface for optical applications. Conventional optical components can be very large as they include items such as convex lenses, phase-retardation plates, polarizers, optical filters and the like. In many cases they cannot be made much smaller as their optical properties will only emerge with a certain thickness or volume of material. In contrast, metasurface based optical components can be made very thin and therefore will be suited to broader range of portable applications where size and weight are key design considerations.

Although there are still many problems to overcome, it is already evident that sub-wavelength nano-structured metasurfaces can be used to avoid many of the issues identified with conventional systems, as will be demonstrated in the following chapters.

In Chap. 3, introduces a new method to increase transmittance in coaxial hole array based filters by tuning localized surface plasmons and surface plasmon polaritons. This overcomes one of the chief problems with earlier such plasmonic color filters.

Chapter 4 demonstrates an experiment to integrate a color filter onto an image sensor to realize CMY color imaging, i.e., that employs the subtractive colors cyan, magenta and yellow. As the optical transmission is much higher through a CMY filter, a CMY camera is better suited to operate in low light conditions than its RGB equivalent. The chapter presents a CMY camera developed using a nanoscale color filter mosaic formed from Al–TiO_2–Al nanorods integrated onto a commercial CMOS image sensor.

---

[4] Note that the array period is given as P in (1.14)–(1.17) and as Λ in Fig. 1.7b.

Multispectral cameras capture images in multiple wavelengths within narrow spectral bands and are well suited to a broad range of applications, such as agro-forestry research, medical analysis and so on. Chapters 5 and 6 propose and analyse a number of novel narrow bandpass filters that employ a hybrid combination of single plasmonic layer and dielectric layers. One of the proposed multispectral filter arrays has been fabricated and integrated on to a commercial image sensor in order to demonstrate its operation on a typical precision agriculture application.

We offer some comments on the future outlook for multispectral image sensors and their applications in the final chapter of the book.

References

1. S. Chang, X. Guo, X. Ni, Optical metasurfaces: progress and applications. Ann. Rev. Mater. Res. **48**(1), 279–302 (2018)
2. R. Mittra, C.H. Chan, T. Cwik, Techniques for analyzing frequency selective surfaces–a review. Proc. IEEE **76**(12), 1593–1615 (1988)
3. J.M. Vaughan, *The Fabry-Pérot Interferometer* (Theory, Practice and Applications (CRC Press, History, 1989)
4. M. Born, E. Wolf, Chapter VII—elements of the theory of interference and interferometers, 6th edn. in *Principles of Optics*, eds. by M. Born, E. Wolf (Pergamon, 1980), pp. 256–369
5. J.E. Harvey, R.N. Pfisterer, Understanding diffraction grating behavior: including conical diffraction and Rayleigh anomalies from transmission gratings. Opt. Eng. **58**(8), 1–21 (2019)
6. W.L. William, *Introduction to Imaging Spectrometers* (SPIE Press, Bellingham, Washington, 1997)
7. D. Schroeder, *Astronomical Optics*, 2nd edn. (Academic Press, Sept, 1999)
8. D. Malacara (ed.), *Geometrical and Instrumental Optics Methods in Experimental Physics*, vol. 25 (Academic Press, 1989)
9. B.E.A. Saleh, M.C. Teich, *Fundamentals of Photonics*, 3rd edn. (John Wiley & Sons, Ltd.)
10. N. Das, D. Chandrasekar, M. Nur-E-Alam, M.M. K Khan, Light reflection loss reduction by nano-structured gratings for highly efficient next-generation gaas solar cells. Energies **13**(16) (2020)
11. O. Boyko, F. Lemarchand, A. Talneau, A.-L. Fehrembach, A. Sentenac, Experimental demonstration of ultrasharp unpolarized filtering by resonant gratings at oblique incidence. J. Opt. Soc. Am. A **26**(3), 676–679 (2009)
12. H. Raether, *Surface Plasmons on Smooth and Rough Surfaces and on Gratings*, vol. 111 (Springer Tracts in Modern Physics, Springer-Verlag, Berlin Heidelberg, 1988)
13. B. Groever, N.A. Rubin, J.P. Balthasar Mueller, R.C. Devlin, F. Capasso, High-efficiency chiral meta-lens. Sci. Rep. **8**(1), 7240 (2018)
14. W.T. Chen, A.Y. Zhu, J. Sisler, Z. Bharwani, F. Capasso, A broadband achromatic polarization-insensitive metalens consisting of anisotropic nanostructures. Nat. Commun. **10**(1), 355 (2019)
15. C. Williams, Y. Montelongo, T.D. Wilkinson, Plasmonic metalens for narrowband dual-focus imaging. Adv. Opt. Mater. **5**(24), 1700811 (2017)
16. S. Wang, P.C. Wu, V. Su, Y.-C. Lai, M.-K. Chen, H.Y. Kuo, B.H. Chen, Y.H. Chen, T.-T. Huang, J.-H. Wang, R.-M. Lin, C.-H. Kuan, T. Li, Z. Wang, S. Zhu, D.P. Tsai, A broadband achromatic metalens in the visible. Nat. Nanotechnol. **13**(3), 227–232 (2018)
17. X. Yin, T. Steinle, L. Huang, T. Taubner, M. Wuttig, T. Zentgraf, H. Giessen, Beam switching and bifocal zoom lensing using active plasmonic metasurfaces. Light Sci. Appl. **6**(7), e17016–e17016 (2017)
18. J. Bai, C. Wang, X. Chen, A. Basiri, C. Wang, Y. Yao, Chip-integrated plasmonic flat optics for mid-infrared full-stokes polarization detection. Photon. Res. **7**(9), 1051–1060 (2019)

19. N.A. Rubin, G. D'Aversa, P. Chevalier, Z. Shi, W.T. Chen, F. Capasso, Matrix Fourier optics enables a compact full-Stokes polarization camera. **365**(6448) (2019)
20. Thorlabs Inc. Thorlabs imaging. https://www.thorlabs.com/navigation.cfm?guide_id=2268. Accessed June 2021
21. Edmund Optics Inc. 45° 355 nm crystalline quartz polarization rotator. https://www.edmundoptics.com.au/p/45deg-355nm-crystalline-quartz-polarization-rotator/3695/. Accessed June 2021
22. J.K. Gansel, M. Thiel, M.S. Rill, M. Decker, K. Bade, V. Saile, G. von Freymann, S. Linden, M. Wegener, Gold Helix photonic metamaterial as broadband circular polarizer. Science **325**(5947), 1513–1515 (2009)
23. J.P. Balthasar Mueller, N.A. Rubin, R.C. Devlin, B. Groever, F. Capasso, Metasurface polarization optics: independent phase control of arbitrary orthogonal states of polarization. Phys. Rev. Lett. **118**, 113901 (2017)
24. T.W. Ebbesen, H.J. Lezec, H.F. Ghaemi, T. Thio, P.A. Wolff, Extraordinary optical transmission through sub-wavelength hole arrays. Nature **391**(6668), 667–669 (1998)
25. J.-G. Yun, S.-J. Kim, H. Yun, K. Lee, J. Sung, J. Kim, Y. Lee, B. Lee, Broadband ultrathin circular polarizer at visible and near-infrared wavelengths using a non-resonant characteristic in helically stacked nano-gratings. Opt. Express **25**(13), 14260–14269 (2017)
26. E. Hutter, J.H. Fendler, Exploitation of localized surface plasmon resonance. Adv. Mater. **16**(19), 1685–1706 (2004)
27. W.L. Barnes, A. Dereux, T.W. Ebbesen, Surface plasmon subwavelength optics. Nature **424**(6950), 824–830 (2003)
28. V.G. Kravets, A.V. Kabashin, W.L. Barnes, A.N. Grigorenko, Plasmonic surface lattice resonances: a review of properties and applications. Chem. Rev. **118**(12), 5912–5951 (2018). (PMID: 29863344)
29. M.F. Limonov, M.V. Rybin, A.N. Poddubny, Y.S. Kivshar, Fano resonances in photonics. Nat. Photon. **11**(9), 543–554 (2017)
30. T.Y. Kang, W. Lee, H. Ahn, D.-M. Shin, C.-S. Kim, J.-W. Oh, D. Kim, K. Kim, Plasmon-coupled whispering gallery modes on nanodisk arrays for signal enhancements. Sci. Rep. **7**(1), 11737 (2017)
31. W. Wei, X. Zhang, Y. Hui, Y. Huang, X. Ren, Plasmonic waveguiding properties of the gap plasmon mode with a dielectric substrate. Photon. Nanostruct. Fundamentals Appl. **11**(3), 279–287 (2013)
32. S. Kasani, K. Curtin, W. Nianqiang, A review of 2D and 3D plasmonic nanostructure array patterns: fabrication, light management and sensing applications. Nanophotonics **8**(12), 2065–2089 (2019)
33. S.A. Maier. *Plasmonics—Fundamentals and Applications*, 1st edn. (Springer, 2007)
34. C. Valsecchi, L.E.G. Armas, J.W. de Menezes, Large area nanohole arrays for sensing fabricated by interference lithography. Sensors **19**(9) (2019)

Chapter 2
Metasurfaces and Multispectral Imaging

Since the first color digital camera was developed in the 20th century, image sensor research has enjoyed a boom time. As it is the key component that converts an optical signal to digital, the image sensor is a critical part of any digital camera. This digital signal can be further processed for color reconstruction, image correction, feature extraction and a myriad of other functions that can be performed on the raw image. Multispectral cameras extend the concept of conventional color cameras to capture images across multiple spectral bands. Images from a multispectral camera can extract significant additional information that the human eye or a normal camera fails to capture. As a result, they have found important applications in fields such as precision agriculture, forestry, medicine, as well as object identification and classification.

A good illustrative example of the use of Multispectral Image Systems (MIS) is the Multispectral Imaging, Detection and Active Reflectance (MiDAR) remote sensing project funded by NASA [1]. The MiDAR system generates high-frame-rate multispectral video in response to coded narrowband structured illumination that can be transmitted from a range of platforms such aircraft, rovers, and submersibles, thereby supporting a range of potential applications including high-resolution dynamic multispectral image analysis from air, space and underwater environments [2].

The normalised difference vegetation index (NDVI) is a common measure in remote sensing for agriculture. As healthy leaves reflect more NIR light than dead or stressed leaves [3], UAVs integrated with MIS can over-fly large regions and very quickly determine the NDVI of a crop based on:

$$NDVI = \frac{(NIR - R)}{(NIR + R)}.\tag{2.1}$$

In this way, it becomes easy to evaluate plant condition in real time and to predict important parameters, such as nutrient requirements, water stress and/or harvest rates.

© The Author(s), under exclusive license to Springer Nature Singapore Pte Ltd. 2021
X. He et al., *Multispectral Image Sensors Using Metasurfaces*, Progress in Optical Science and Photonics 17, https://doi.org/10.1007/978-981-16-7515-7_2

MIS also plays an important role in the healthcare area. MIS techniques operating in red end of the visible and NIR are currently used in hospitals to locate the position of veins [4]. As NIR light can penetrate the skin and increase the contrast between blood and skin, images of the vein under the skin can be easily found and superimposed on top of the skin. It also can detect disease within some structures in the body [5] such as distinguishing ulcers and erythematous regions.

This chapter will begin by discussing the general concepts that underpin the application of metasurfaces to imaging systems, some of which were introduced in Chap. 1. The basic structure of a pixel will be described along with some basic color image sensor techniques on which the idea of a multispectral camera is based. To achieve multispectral imaging, various optical devices with different mechanisms and photonic theories have been presented in both an industrial and academic context. For example, early conventional multispectral cameras were made up of multiple image sensors each externally fitted with a narrow passband wavelength filters, optics and multiple electronics. A small number of commercial products that use this approach, such as Tetracam, Red Edge etc., will be discussed.

The need for multiple sensors to cover the bands in a multispectral camera results in a number of problems. These systems can be bulky, power hungry and may suffer from image co-registration problems which in turn limits their wider usage. These problems can be eliminated if the camera is based on a single image sensor. There are fundamentally two ways to achieve this objective. The first is to arrange all the narrow passband filters in an array which is then integrated onto the sensor surface. This approach is employed by companies such as IMEC, Spectral Devices and others, as will be discussed below. Each narrow passband filter may have its own recipe based on Fabry-Pérot theory, and so this type of multispectral camera requires overlay and alignment processes for every filter, which can become difficult and expensive. The ideal filter would be able to be fabricated in either a single or small number of steps on a single layer of material. This approach will be introduced here and expanded further in later chapters.

2.1   Conventional Color Filters and Image Sensors

Digital image acquisition by an image array is a straightforward sampling process so its resolution will depend directly on the number of sampling elements-pixels— available in the array. In a typical case, such as that shown in Fig. 2.1, a micro-lens array is positioned on top in order to focus the incident light [6–15]. The focused light will then reach the color filter array (CFA), and only the desired color of light, i.e., within a particular band of wavelengths, will pass through this layer to be collected by the photodetector (PD). Subsequent steps may include additional analog processing or just analog to digital conversion [6–18], followed by color reconstruction and the application of interpolation, or smoothing algorithms (also called "demosaicing"). Additionally, the black matrix within the structure shown in Fig. 2.1 acts to isolate adjacent pixels when integrating the CFA onto a substrate.

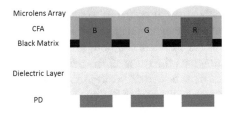

Fig. 2.1 Simplified 2-D pixel structure comprising an aligned microlens and color filter array (CFA) mounted over a photodiode (PD) array separated by a dielectric layer. The opaque black matrix helps to reduce crosstalk between the PD sensors due to the thickness of the dielectric layer

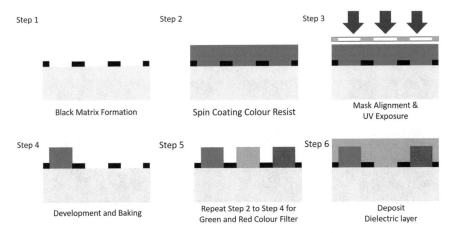

Fig. 2.2 Manufacturing process for fabricating conventional color filter on a thin dielectric substrate illustrating the various alignment steps required to correctly position the CFA

One example of the integration of CFA onto an image sensor can be found in [13], and an example of the process is illustrated in Fig. 2.2. In this case a very thin dielectric, typically Si_3N_4 or SiO_2 is used to form the substrate, which is in direct contact with the PD within the image sensor. A thin dielectric substrate is desirable as it results in less spatial crosstalk between adjacent pixels. Ideally therefore, the CFA should directly contact the PD. There are many different kinds of conventional color filter materials in use, most of them company-confidential. We will use an organic polymer-based color resist here as a general example of the process. This color filter material can be spin-coated onto a substrate and behaves like a negative photo—resist in that the part exposed by the UV light becomes harder and will remain after development and baking. Both the exposure and development times must be precisely controlled as exposure times that are too long or short can result in shape distortion, particularly at the corners. Development time must also be tightly controlled, preferably within a one second tolerance, as the time directly impacts the size of the developed area and therefore the accuracy of the shape boundaries. Similarly, the baking temperature must be carefully monitored as the color resist material can melt or de-laminate from the substrate at excess temperatures. Further, the color resist material itself can be sensitive to the UV radiation used to expose the resist as well as some of the chemical solutions used as a developer. The same steps need to be repeated numerous times for all the other filter stages. Thus, during the UV exposure the masks must be carefully aligned, both with the underlying PD and with their adjacent layers. Finally, a thin dielectric layer is deposited resulting in a flat surface on which the micro-lens array can be integrated. In some of the commercial work that has been disclosed, ITO is used for this layer [13].

We have undertaken a brief experiment to illustrate the structure and properties of a typical CFA and micro-lens array. Here, the image sensor of a mid-range commercial camera (CANON 1000D) was used by removing the sensor after disassembling the camera (Fig. 2.3). The process involved first removing the protective glass from the sensor using the following steps:

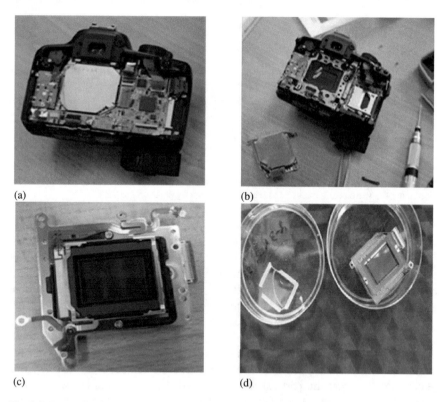

(a)   (b)

(c)   (d)

Fig. 2.3 Removing the protective glass from image sensor. **a** CANON 1000D camera rear view with PCB boards covering the image sensor disassembled from main body. **b** Image sensor assembly removed. **c** Top view of the image sensor showing (top to bottom) the low pass filter (UV blocking filter) and protective glass. **d** Image sensor shown with glass removed

- The PCB boards covering the image sensor were disassembled from the main body of the camera (Fig. 2.3a);
- The image sensor assembly was carefully removed using tweezers (Fig. 2.3b–c);
- The low pass filter (UV blocking filter) was removed with a knife;
- Finally, the protective glass was heated (keeping the temperature below 80 °C) using a 2W laser cutter and lifted off with a knife (Fig. 2.3d).

Following its removal from the camera, the next step was to remove the Bayer layer from the image sensor. Firstly, the sensor was exposed to a 50W oxygen plasma for two hours, which softened the micro-lens array and color filter layers sufficiently to allow them to be easily removed with a wooden scraper (Fig. 2.4a). In the microscope image shown in Fig. 2.4b, the micro-lens array and CFA are totally removed on the left side, while only the micro-lens array was removed from the area in the center of the image.

In Fig. 2.4c, a scanning electron microscopy (FEI NovaNanoSEM-430) was used to image the PD under each pixel more accurately. It was observed that the sensitive area is about 70% of the overall pixel size of 5.6 µm × 5.6µm indicating that in order to avoid introducing spatial crosstalk, the tolerance for misalignment between the CFA and PD must be less than 1.7 µm. In Fig. 2.4d, an optical profilometer (Bruker Contour GT-I) was used to measure the thickness of CFA, which was found to be approximately 1.5 µm.

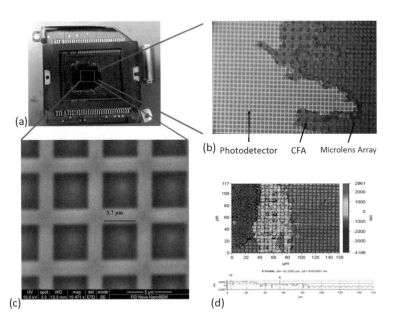

Fig. 2.4 The experimental measurement of CANON 1000D image sensor: **a** after Bayer layer removal; **b** microscopic image of the micro-lens array, CFA and PD; **c** PD measurement with SEM; **d** height measurement with optical profilometry

Conventional color filters have been found to have a number of disadvantages [19–27] in that their materials (or at least the ones that have been disclosed to date) appear to be sensitive to UV light as well as some chemical solutions, such as developer. They can also become unstable at elevated temperatures. Further, the organic dye-based material requires a minimum volume in order to produce different scattering colors, which results in a large filter thickness (1.5 μm in our experiment, above). This has the potential to introduce more spatial crosstalk between adjacent pixels when their size is reduced to submicron scale. Finally, CFA manufacture requires multiple, carefully aligned steps for each filter, which increases the fabrication cost and complication as the number of filters increases.

2.2   Existing Multispectral Imaging Systems

The concept of a Multispectral Imaging System (MIS), which captures images in narrow bands across the electromagnetic spectrum, was invented decades ago and since then has attracted much research across a range of applications. Depending on the application, the spectral width can vary between 10 and 90 nm. In the next few sections, we demonstrate three types of existing MIS devices and briefly highlight their advantages and disadvantages.

2.2.1   Subjective MIS Using Narrow-Band Illumination Sources

One of the easiest ways to capture images with multiple spectral bands is by sequentially illuminating the target object with a narrow-band light source. This is also known as a subjective imaging system as the illumination source is subjectively held in front of the objects that will be captured by the camera. Conversely, objective imaging means there are no modifications made to the illumination and the object is directly captured by a camera under normal lighting conditions (which could be natural sun-light). While subjective multispectral imaging systems now tend to use commercially available narrow-band LEDs as an illumination source, there are other ways to extract the desired narrow passband light source from a general white light source. For example, prisms or beam splitters have been previously used in spectrometer systems and can also be used for MIS [2]. As an example of this, in Fig. 2.5 a light source is split into three color channels (bands) using two mirrors, achieving a FWHM of around 100 nm. Although each channel is captured at the full resolution of the camera and therefore has, by definition, N times the resolution of an objective MIS with the same camera (N is the number of bands), this kind of image system with its complicated optics can be quite bulky, particularly when large numbers of bands

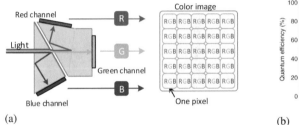 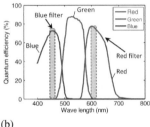

(a)   (b)

Fig. 2.5 a Schematic diagrams for narrow passband illumination sources, **b** quantum efficiency curves. From [2], ©Springer Nature, reproduced with permission

are required. One example described in [28], called the *Multispectrum Pushbroom* airborne camera system is nearly 900 mm in length and weighs just under 35 kg.

2.2.2 Objective MIS Using Random Optical Filters

Using random optical filter is an alternative way for spectrometers to use prisms or beam splitters to select desired narrow-band sources. The working principles of both methods are based on the general sampling theory:

$$S_i = \int I(\lambda)T_i(\lambda)\mu(\lambda)d\lambda, \tag{2.2}$$

where S_i is the measured signal under the ith illumination source or random filter, $I(\lambda)$ is the spectrum to be measured, $T_i(\lambda)$ is the transmission spectrum of ith source or random filter, and $\mu(\lambda)$ refers to the quantum efficiency of the detector.

Because we are now sampling multiple narrow-band channels, the number of samples to be processed increases proportionally. While reconstructing the spectrum with multiple signal measurements under narrow-band sources is quite straightforward, there is an obvious tradeoff between the resolution of the recovered spectrum, which depends on the FWHM of the light source, and the overall time taken to derive the result. To derive an accurate representation of the target spectrum, the samples typically have to be "normalised" to a baseline given by the response to broad-band ('white') illumination. The objective here is to remove the errors introduced by variability in both the filter transfer functions and the underlying quantum efficiency of detectors. Various approaches have been proposed to estimate the transmission functions that are required recover the spectrum in the random filter approach. These include capturing global information about the signal spectrum [29] plus a range of compressive sensing methods aimed at reducing the error between the measured and target spectra [3, 30–35].

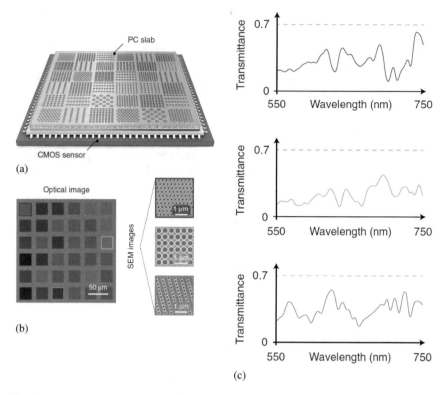

Fig. 2.6 Micro-spectrometer based on photonic-crystal slabs. From [30], licensed under a creative commons attribution (CC BY 2.0)

Random spectral filters based on photonic crystals (PC) have been described in [30]. The technique used was to mill circular nanoholes with various orientations and sizes into a thin silicon layer to make 36 broad random filters, as shown in Fig. 2.6. The operating wavelength of this PC spectrometer is from 550 nm to 750 nm, A one snap-shot ultra-spectral camera was also developed later with a similar working theory [35]. In that case, a number of different geometries were patterned into a silicon film to fabricate 25 random filters and the device was then used to reconstruct spectral images under different bands across transmission spectrum of the test objects. It is worth noting that, because silicon has a large optical loss in the blue region below 450 nm, neither of these proposals capture much spectral information in that band. Research is still ongoing and it is early days yet, but it appears reasonable to expect that random optical filters might eventually offer a useful alternative to narrow-band filters in multi-spectral cameras. Ultimately, applying some form of compressive sampling technique or the type of deep neural network illustrated in Fig. 2.7 [36] could lead to a reduction in the required number of filters, thereby simplifying the overall design of these multi-band systems.

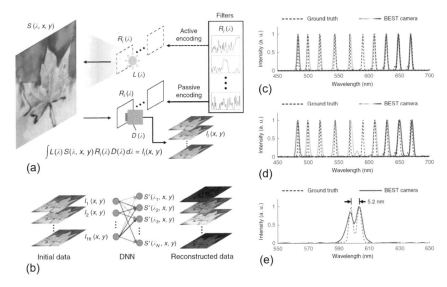

Fig. 2.7 Schematic diagram of the deep neural network used for spectral camera based on 16 random filters. From [36], licensed under a creative commons attribution (CC BY 2.0)

2.2.3 Objective MIS Using Multiple Cameras and Optical Filters

This type of MIS merges separate bandpass filters with multiple cameras [37, 38]. One key advantage here is that the image within each band is captured at full resolution. However, it still has a few drawbacks, the main one being that the multiple optics and sensors increase its overall size and and weight, which will limit its application in some cases. In addition, as these multiple camera systems take each of the individual images from a slightly different perspective, these will typically require further processing to reduce or remove image co-registration problems [39–42].

Figure 2.8 illustrates a MIS based on a simple optical filter wheel with a single camera. The optical filters of different bands can be changed by a motor and control system so images in each spectrum are still captured at full camera resolution. However, this MIS exhibits similar problems to the previous system. The narrow band images corresponding to a single "shot" are separated in time, thus causing problems with overlapping moving targets and the like. The overall system is still potentially bulky and power hungry.

Fig. 2.8 Diagram of spectral camera using optical filter wheel. From [5], licensed under a creative commons attribution (CC BY 2.0)

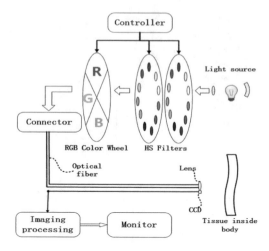

2.2.4 MIS Using Integration of Filter Array with Single Image Sensor

Single sensor based multispectral image systems have emerged recently onto the market [43–45]. This class of MIS integrates the narrow bandpass filter array directly on the image sensor. However, while in most cases the details of the specific filter technology are carefully guarded trade secrets, some narrow bandpass filter patents plus work in the public domain from IMEC [43] reveal filter structures based on multilayer coatings. In this case, the peak wavelength can be tuned by varying the material thickness. As a result, each filter has a unique thickness, and the fabrication process required to integrate the filter array onto the image sensor would be similar to the conventional color image sensor technology. The number of processing steps is directly proportional to the number of bands, which will significantly increase the fabrication cost and complexity for multi-spectral systems, particularly those with a large number of bands.

2.3 Existing Narrow Passband Filter Techniques

This section will focus on various optical filter designs with both dielectric-based (e.g., prism, multilayer coating, dielectric guided-mode resonant filters) and metal-based structures (e.g., plasmonic filters).

2.3.1 Multilayer Coatings

The basic structure of a multilayer coating structure is based on the Fabry-Pérot (FP) etalon, which was described previously in Sect. 1.1.1. The mirrored surfaces in

Fig. 1.3 (Chap. 1) can be replaced by other metal dielectric materials with a thickness of $\lambda/4n$, where n is the refractive index of the layer, that act as anti-reflective layers. Many metal-dielectric-metal (MDM) resonators have been proposed to reduce reflections have also been used for color filtering [46], as absorber layers [47–49] and anti-reflection coatings [50–53].

As just mentioned, the mirror does not need to be metallic as a dielectric boundary can also act as a mirror. In this case, a large refractive index difference between the two adjacent layers is the key factor that enables the filters to be produced with narrow bandwidths. In [54], the FP "mirror" is made of combinations of dielectric materials with larger refractive index material (i.e., TiO_2) surrounding a smaller one (i.e., SiO_2 or MgO). The resulting FWHM was 10 nm indicating that this kind of structure can be considered to be a narrow-band filter. An alternative structure presented in [55] is based on a double FP with a silver mirror and a cavity formed from Ta_2O_5, which results in a FWHM of 50 nm. Further, by replacing the center dielectric with photoresist, the FWHM could be reduced to less than 50 nm [56].

A multilayer film structure based on one-dimensional photonic crystals was proposed and analyzed in [57, 58]. The propagation speed of light will change in different dielectrics according to the equation: $ck = \omega(k)n$, where n is the refractive index of dielectric, k is the wavevector that is periodic in the 1-D photonic crystal structure with a period of $2\pi/P$, P being the period of the multilayer. When light propagates through this multilayer structure, two adjacent layers with different refractive indices will create a photonic band gap. Note that $\Delta\omega$, the mode with the frequency in the bandgap, will be rejected and therefore light with the corresponding wavelength will be rejected. The gap to mid-gap ratio can be approximated as:

$$\frac{\Delta\omega}{\omega_m} \approx \frac{\sin(\frac{\pi l_1}{P})\Delta n^2}{\pi n^2}, \tag{2.3}$$

where ω_m is the mid-gap frequency and n is the smaller refractive index of the adjacent two layers. It is clear that this ratio will reach maximum when:

$$\frac{l_1}{P} = \frac{1}{2}. \tag{2.4}$$

The bandgap will increase as the refractive index difference becomes larger. In other words, if layer 1 has the refractive index n_1 and thickness l_1 and its adjacent layer 2 with n_2 and l_2, then when $l_1 n_1 = l_2 n_2$, the gap size will be maximized and therefore the incident light will produce a very narrow spectrum. Such an effect can be used to create very narrow optical filters.

Normally, photonic crystal structures are analysed by considered all layers as being formed from dielectric materials. Prior proposals [57–59] in which the adjacent layers are made up of pairs of Si/SiO_2 layers repeated twelve times have shown significantly reduced values of FWHM. In [60], the Metallic Photonic Bandgap structure is repeated 40 times to further reduce the FWHM. A large number of patents relating to narrow bandpass filters with multilayer structures already exist (e.g., [61–65]).

2.3.2   *Dielectric Guided-Mode-Resonance Filters (GMRF)*

The photonic bandgap filter is based on a dielectric multilayer repeating in the vertical direction. In contrast, the unit nanostructure of the dielectric GMRF repeats in the horizontal direction shown in the Fig. 2.9. As shown, the permittivity of the top layer and the spacing around the grating (medium 1) is ε_1, while ε_2 and ε_3 are the permittivity of the grating and substrate (medium 2 and 3), respectively.

In this GMRF:

$$\varepsilon_2 = \varepsilon_g + A\cos(\frac{2\pi}{p}x), \tag{2.5}$$

where A is the amplitude and p is the pitch (period) of the grating [66]. Therefore, the condition of the GMRF is given by (2.6):

$$\varepsilon_2 > \varepsilon_g > \varepsilon_1, \varepsilon_3, \tag{2.6}$$

In [68], a single SiN grating layer forming a 1-D GMRF was demonstrated with a FWHM of less than 1 nm, while in [69], a double grating structure was shown to decrease the FWHM. However, these techniques may only be able to suppress secondary peaks and produce a sharp peak at certain wavelengths within a short range (less than 10 nm). By sealing the dielectric grating with another material (e.g., SiO$_2$, as shown in Fig. 2.10, and also increasing the refractive index difference between nH

Fig. 2.9 Simplified 2-D structure of the GMRF in [66]

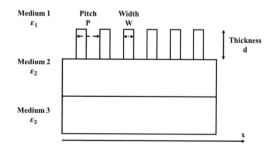

Fig. 2.10 Improved dielectric GMRF structure in [67]

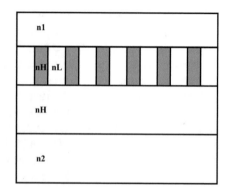

and nL [67], it is possible to achieve a FWHM of less than 1 nm as well as suppressing other peaks across the visible wavelength range. This structure will become more effective when the peak wavelength can be more easily tuned. Further, 2-D GMRFs have also been proposed as good candidates for narrow band filters [70–72]. However, these kinds of filters only work in the reflection mode and at some specific incident angles in transmission mode and thus are not generally suited for use in multispectral filtering.

2.3.3 Silicon Nanowire

The optical properties of dielectric nanowires will depend on their radius r and the refractive index n of the material with the relationship [73]:

$$\lambda \approx \frac{2\pi rn}{p}, \qquad (2.7)$$

for integer p = 1,2,3, …etc. Equation (2.7) can also be used for dielectric nano-rod, resonator or nano-disk structures, even at larger sizes. In fact, it may even be valid when the dielectric material with large refractive index exhibits high confinement modes. Silicon has a large refractive index (as high as 4), and is therefore suitable for use in color filtering [73–78]. For example, a multispectral filter based on Si nanowire was presented in [75] and used to perform multispectral imaging. A color image was able to be reconstructed from the multispectral camera based on the integration of the Si nanowire array onto a monochrome image sensor and, by using NIR illumination, objects behind a black screen could be observed by analyzing the data from NIR band of the multispectral camera [75] (Fig. 2.11).

2.4 Plasmonic Filters Based on Metallic Surfaces

As introduced in Sect. 1.2.3, it has been observed over many years that light will be diffracted while passing through small metallic hole less than 100 nm in diameter [79, 80] and that this diffraction will be wavelength dependant. The holes may be circular, rectangular, even non-regular. Regardless of their shape, the behavior of the surface can be described and predicted in terms of plasmonic theory, which comprises two cases:

- surface plasmon polaritons (SPP) occur if the plasmon is excited and propagated at the interface between the metal and dielectric;
- localized surface plasmons (LSP) are evident when the excited plasmons are not propagating at the surface of the metallic nanoparticle (e.g. nano-sphere, nano-disk) but are still coupling to the electromagnetic field [81].

In the following sections, we briefly examine a small number of multispectral filter topologies based on metallic nano-particles and thin metasurfaces.

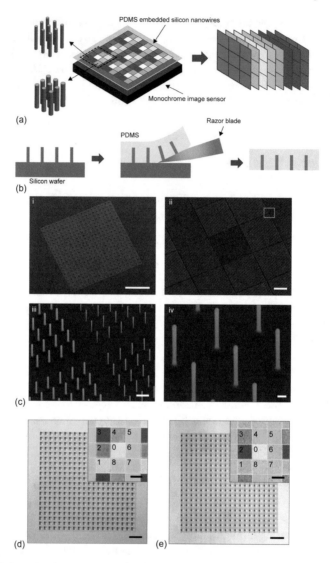

Fig. 2.11 Schematic of nanowire based multispectral filter showing **a** basic concept; **b** manufacturing process using a PDMS template; **c** SEM image of array of etched vertical silicon nanowire array showing 20 × 20 unit cell organisation; **d** reflection-mode optical microscope image of etched vertical silicon nanowire array on silicon substrate; **e** transmission-mode optical microscope image of PDMS-embedded vertical silicon nanowire array. From [75], ©Springer Nature, reproduced with permission

2.4.1 Nano-Slit Gratings

The propagation constant for metallic nano-slit grating structures, considering only the first scattering order, is give by:

$$\beta = \pm \frac{2\pi}{P} + k_0 \sin \theta, \qquad (2.8)$$

where P is the period of the grating and θ is the angle of incident of the light. When the incidence angle is normal to the sensor surface, $\theta = 0$ so that:

$$k_0 \sqrt{\frac{\varepsilon_1 \varepsilon_2}{\varepsilon_1 + \varepsilon_2}} = \frac{2\pi}{P}, \lambda = P \sqrt{\frac{\varepsilon_1 \varepsilon_2}{\varepsilon_1 + \varepsilon_2}}. \qquad (2.9)$$

Thus, it can be seen that the peak wavelength is proportional to the grating period. The influence on the optical transmission efficiency with grating thickness has been investigated [82] using this theoretical framework and many color filters based on metallic grating have been proposed and analyzed. For example, in [26], a 1-D Ag nano-slit grating on quartz was shown to reach 60% transmission efficiency while producing subtractive color primaries: yellow, magenta and cyan. More recently, a MDM nano-slit grating was shown to produce an additive color gamut, i.e., red, green and blue [83, 84]. Metal nano-slit gratings have also been used for the narrow bandpass filters with smaller FWHM [27, 85–87].

The plasmonic filter shown in Fig. 2.12 is formed from an array of stacked MDM strips which repeat in the horizontal direction forming a regular pattern of nano-slits, exhibits a FWHM of 64 nm at a wavelength of around 1500 nm [27]. The Ag/SiO$_2$ MDMDM stack in [85] (Fig. 2.13a),[1] shows a similar FWHM to the optimised MDM array (less than 30 nm at around 680 nm), but greatly suppresses the secondary peak that appears at smaller wavelengths (Fig. 2.13b). Finally, the simple Al nanoslit formed on top of a Al$_2$O$_3$ buffer layer on a glass substrate in [87] (Fig. 2.14a) operates with interacting guided-mode and surface-plasmon resonance effects near the Rayleigh anomaly wavelength and exhibits a greatly reduced FWHM, down to around 14nm in the experiments of [87] (Fig. 2.14b).

However, a major problem with metallic nano-slit gratings is that they are polarization dependent so that the spectrum under a TE wave will be different from that under a TM wave, something that has been proved from both theory and experiments over many years [19–27, 81, 88, 89]. This feature may be a disadvantage for imaging applications if the incident light is polarized and thus where the optical system needs additional optics, such as a polarizer. When the incident light combines TE and TM wave types, and one polarization mode is blocked (so that only one is effective) the available optical power is halved.

---

[1] Called Metal-Insulator-Metal (MIM) in [85].

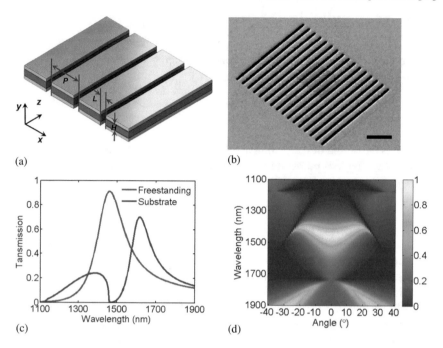

(a)

(b)

(c)

(d)

Fig. 2.12 Plasmonic bandpass filter formed from an array of stacked metal-dielectric-metal (MDM) structures **a** schematic of the MDM array with a Si_3N_4 dielectric layer (blue shading) sandwiched in between Au layers, **b** SEM image of the device (with 4 μm scale bar). **c** Transmission spectra of the device: freestanding versus on substrate. **d** Simulated transmission diagrams as a function of incident angle and wavelength for the free-standing array. From [27], licensed under a creative commons attribution (CC BY 2.0)

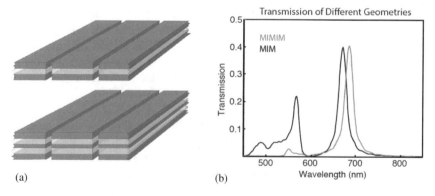

(a)

(b)

Fig. 2.13 MDM vs. MDMDM nano-slit gratings **a** Simplified schematic of the arrays. **b** Relative transmission results showing the improved suppression of out-of-band transmission in the MDMDM device. From [85], licensed under a creative commons attribution (CC BY 2.0)

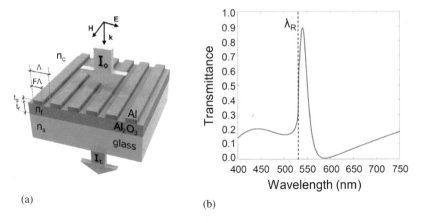

(a)   (b)

Fig. 2.14 a Simplified schematic of multispectral filter based on Al nano-slit array on an Al$_2$O$_3$ buffer layer on glass. **b** Normalized transmittance of green wavelengths (array period Λ = 350 nm) at a fixed Al$_2$O$_3$ film thickness of 200 nm, illustrating the narrow ($\sim$14 nm) FWHM. The blue and red cases (Λ = 290 and 430 nm, respectively) exhibit similar FWHM. The vertical dashed line indicates the Rayleigh anomaly wavelength λ_R. From [87], ©The Optical Society (United States), reproduced with permission

2.4.2 Nanohole Arrays

There are two common arrangements of metallic nanohole array: square and hexagonal. The propagation constant (β) of both of these topologies can be described as [80]:

$$\beta = K_0 \sin\theta \pm nG_x + nG_y = K_0 \sin\theta \pm (n+m)\frac{2\pi}{P} \qquad (2.10)$$

and the peak wavelength (λ) will be given for a square array by:

$$\lambda = \frac{P}{\sqrt{\frac{4}{3}(n^2+m^2)}}\sqrt{\frac{\varepsilon_1\varepsilon_2}{\varepsilon_1+\varepsilon_2}}, \qquad (2.11)$$

and a hexagonal hole array by:

$$\lambda = \frac{P}{\sqrt{\frac{4}{3}(n^2+nm+m^2)}}\sqrt{\frac{\varepsilon_1\varepsilon_2}{\varepsilon_1+\varepsilon_2}}, \qquad (2.12)$$

where n and m are the scattering order, ε_1 and ε_2 are the dielectric constant of the dielectric and metal.

These models have been used previously to analyze the behavior of plasmonic color filters formed from hexagonal hole arrays in aluminum films [19, 23]. The FWHM has been shown to vary from 90 nm to 180 nm while the peak wavelength is within the red region. Moreover, a plasmonic color filter formed using a hexagonal hole arrangement is polarization independent and has higher transmission efficiency than the square array [23].

Recently, a large plasmonic color filter array was fabricated and integrated on a monochrome CMOS image sensor. After the image processing steps such as demosaicing, white balance and gamma correction, a full color image could be reconstructed [90]. Multispectral filters based on metallic hole arrays have been presented that operate at visible and NIR wavelengths [22, 91, 92]. A nanohole array inserted in gold film was fabricated with EBL patterning followed by wet etching [92]. The measured FHWM is around 100 nm with 60% transmission efficiency. This multispectral filter has been integrated with an optical system, including a telecentric lens and CMOS image sensor, to image leaves at different wavelengths. Another multispectral filter based on a gold nanohole array was fabricated on a GaAs substrate for the mid IR imaging from 7 to 14 µm. It is assembled with a thermal camera in a black body, to reconstruct images of hot metal objects under 100 °C [92]. The FWHM is around 450 nm.

2.4.3   Nanoparticles

The optical properties of metallic nanoparticles can be described in term of Localized Surface Plasmons (LSP), in which the resonant wavelength is strongly determined by its size, rather than the period [81, 93–98]. Note that the resonant wavelength here is the valley wavelength, the region of minimum transmission. The resulting color is subtractive, such as cyan, magenta and yellow, rather than additive as in red, green and blue. Examples includes arrays using aluminium disks [94, 95], silver disks [96] and gold nanorods [98]. The 40 nm Ag disk color filter demonstrated in [96] has a maximum transmission efficiency of around 60%, and its resonant wavelength remains constant across a range of incident angles up to about 80°. Even though the first peak is somewhat suppressed, this has little effect on the resulting color. Even the spectrum of the 20 nm Al disk in [94, 95] shows no changes as the incident angle varies from 0° to 70°. The reason is probably that the thinner disk is more symmetrical than the thicker disk and therefore more insensitive to the angle of incidence. Therefore, most LSP based structures are polarization independent and incident angle insensitive, except some structures with larger thicknesses such as nanorod, which offers an advantage over SPP based filters. Moreover, they also transfer much larger optical power levels than the additive color filter, making them better suited for the photographic targets with extremely dark background, such as astronomical imaging. However, there have been no subtractive color filters integrated on the image sensors for color imaging introduced previously, and this will be the subject of Chap. 4.

2.4.4 Apertures

The theory of a single coaxial hole inserted in a metal film has been derived in spherical coordinates [99] and it has been shown that the peak wavelength is determined entirely by the spacing in the coaxial holes, with no effect from the hole size or period. The wavelength dependent propagation constant is:

$$\beta = \frac{m\pi - \varphi}{d},\qquad(2.13)$$

where m = 1, 2, 3, …etc., φ is the reflection phase, d is the thickness of the metal film [24].

Coaxial hole structures also exhibit the Metal-Dielectric-Metal FP effect, and so will also exhibit a narrow spectrum. In [99], a filter was presented where the peak wavelength varies from 500 to 650 nm, as the spacing between Au coaxial hole changes from 20 to 40 nm. The minimum FWHM observed was 40 nm. However, the maximum transmission efficiency of this kind of structure is usually smaller than 20% because of the small spacing [99–101]. The plasmonic coaxial holes also have high tolerance to incident angle (FFOV up to 80°) and the array is polarization independent. We return to this issue of small transmission efficiency with a potential solution [21] in the next chapter .

2.5 Summary

Color filters with various structures and materials have been previously described for which the peak wavelength can be tuned to pass different colors by changing thickness of each layer. These include Fabry-Pérot based filters [46–55], photonic bandgap filters [56–59], and dielectric multilayers [54, 61–65]. In some of the periodic arrays, the peak wavelength is proportional to the pitch size (e.g. dielectric GMRF [66–72], plasmonic grating [26, 27, 82–87] and nanohole array [19, 20, 22, 23, 25, 90–92, 102–104]). For some metallic nanoparticles (e.g., silicon nanowire [73–78], nanodisks [93–96], nanorod [97], nano-rectangles [98]), their sizes will determine the peak wavelength, while in other metallic apertures, the peak wavelength relates to the spacing (e.g., coaxial hole [99–101]). However, the maximum transmission efficiency of this kind of structure is usually smaller than 20% because of the small spacing.

Plasmonic coaxial holes exhibit consistent performance over a range of incident angles (FFOV up to 80°) and the array is polarization independent. To compensate for the disadvantage of the small transmission efficiency, we provides a solution in the next chapter [21]. Different strategies have been employed to reduce the spectral width (FWHM) of the narrow bandpass filters were presented. Most early proposals were based on the dielectric GMRF, for which FWHM values as low than 1 nm were shown. However, this technique covers only a limited range of wavelengths and the peak wavelength cannot be easily tuned [66–69]. Later, multilayer coating technologies with tens of layers were introduced [54, 61–65]. The main advantages of this technique is that the peak wavelength can be finely tuned by changing the thickness. While this can be achieved precisely using atomic layer deposition (ALD), it requires multiple steps for a large multispectral filter array and extremely accurate alignment between stages during fabrication.

The next stage in the development of multispectral imaging saw the introduction of narrow bandpass filters formed from silicon nanowires integrated onto a monochrome image sensor. The primary disadvantage of this technique is its large FWHM, larger than 90 nm in the proposals to date [75]. Metallic nano-slit grating filters have been shown to have a FWHM of around 20 nm [27, 87], but may require additional optics to block one polarization component of the incident light, which then halves the incident optical power. Moreover, to date there has been only one report [105] describing the use of a nano-slit grating in front of an optical system. In this case, the filter was not closely integrated with the image sensor and required very complicated software post-processing to create the image.

A metallic nanohole based narrow bandpass filter was analyzed using simulation in [22] and was predicted have a FWHM less than 50 nm. Further, narrow metallic nanohole arrays have been fabricated and integrated with a bulky optical system to perform the imaging. These were shown to have FWHM values around 100 nm in the visible and NIR wavelength [91] and 400 nm at mid-IR wavelengths [92]. At this point in time there has been no prior work demonstrating the integration of multispectral filter arrays using a single exposure fabrication methodology onto a standard image sensor to perform multispectral imaging. This is the focus of Chap. 5 in this book.

References

1. National Aeronautics and Space Administration (Pg. Ed.: Yael Kovo). Multispectral Imaging, Detection and Active Reflectance (MiDAR). https://www.nasa.gov/centers/ames/cct/office/cif/ved-chirayath. Accessed Sept 2021
2. L. Yu, B. Pan, Color stereo-digital image correlation method using a single 3CCD color camera. Exper. Mech. **57**(4), 649–657 (2017)
3. H. Arguello, G.R. Arce, Colored coded aperture design by concentration of measure in compressive spectral imaging. IEEE Trans. Image Process. **23**(4), 1896–1908 (2014)
4. C.-T. Pan, M. D. Francisco, C.-K. Yen, S.-Y. Wang, and Y.-L. Shiue, Vein pattern locating technology for cannulation: a review of the low-cost vein finder prototypes utilizing near infrared (NIR) light to improve peripheral subcutaneous vein selection for phlebotomy. Sensors **19**(3573) (2019)
5. L. Yao, G. Xiaozhou, Y. Zhong, Z. Han, Q. Shi, F. Ye, C. Liu, X. Wang, T. Xie, Image enhancement based on in vivo hyperspectral gastroscopic images: a case study. J. Biomed. Opt. **21**(10), 101412 (2016)
6. Y. Huo, C.C. Fesenmaier, P.B. Catrysse, Microlens performance limits in sub-2 μm pixel CMOS image sensors. Opt. Express **18**(6), 5861–5872 (2010)
7. W.-G. Lee, J.-S. Kim, H.-J. Kim, S.-Y. Kim, S.-B. Hwang, J.-G. Lee, Two-dimensional optical simulation on a visible ray passing through inter-metal dielectric layers of CMOS image sensor device. J. Korean Phys. Soc. **47**(3), S434–S439 (2005)

8. J.C. Ahn, C.-R. Moon, B. Kim, K. Lee, Y. Kim, M. Lim, W. Lee, H. Park, K. Moon, J. Yoo, Y.J. Lee, B. Park, S. Jung, J. Lee, T.-H. Lee, Y.K. Lee, J. Jung, J.-H. Kim, T.-C. Kim, H. Cho, D. Lee, Y. Lee, Advanced image sensor technology for pixel scaling down toward 1.0 μm (invited). IEEE Int. Electron Dev. Meeting, 1–4 (2008)
9. C.-R. Moon, J.-C. Shin, J. Kim, Y.K. Lee, Y.-J. Cho, Y.-Y. Yu, S.-H. Hwang, D.-C. Park, B.J. Park, H.-Y. Kim, S.-H. Lee, J. Jung, S.-H. Cho, K. Lee, K. Koh, D. Lee, K. Kim, Dedicated process architecture and the characteristics of 1.4 μm pixel CMOS image sensor with 8M density, in *2007 IEEE Symposium on VLSI Technology*, 2007, pp. 62–63
10. P.B. Catrysse, B.A. Wandell, Optical efficiency of image sensor pixels. J. Opt. Soc. Am. A Opt. Image Sci. Vis. **19**(8), 1610–20 (2002)
11. C.C. Fesenmaier, Y. Huo, P.B. Catrysse, Optical confinement methods for continued scaling of CMOS image sensor pixels. Opt. Express **16**(25), 20457–20470 (2008)
12. P.B. Catrysse, B.A. Wandell, Integrated color pixels in 0.18-μm complementary metal oxide semiconductor technology. J. Opt. Soc. Am. A **20**(12), 2293–2306 (2003)
13. Toyo Visual, Structure of color filters. https://www.toyo-visual.com/en/products/fpdcf/colorfilter.html. Accessed June 2021
14. International image sensor society index of past workshops. Technology of color filter materials for image sensor. https://www.imagesensors.org/PastWorkshops/2011Workshop/2011Papers/I01_Taguchi_ColorFilter.pdf. Accessed May 2021
15. G. Agranov, V. Berezin, R.H. Tsai, Crosstalk and microlens study in a color CMOS image sensor. IEEE Trans. Electron Dev. **50**(1), 4–11 (2003)
16. H. Rhodes, G. Agranov, C. Hong, U. Boettiger, R. Mauritzson, J. Ladd, I. Karasev, J. McKee, E. Jenkins, W. Quinlin, I. Patrick, J. Li, X. Fan, R. Panicacci, S. Smith, C. Mouli, J. Bruce, CMOS imager technology shrinks and image performance, in. IEEE Workshop on Microelectronics and Electron Devices **2004**, 7–18 (2004)
17. T.H. Hsu, Y.K. Fang, C.Y. Lin, S.F. Chen, C.S. Lin, D.N. Yaung, S.G. Wuu, H.C. Chien, C.H. Tseng, J.S. Lin, C.S. Wang, Light guide for pixel crosstalk improvement in deep submicron CMOS image sensor. IEEE Electron Dev. Lett. **25**(1), 22–24 (2004)
18. S.-J. Kim, S.-W. Han, B. Kang, K. Lee, J.D.K. Kim, C.-Y. Kim, A three-dimensional time-of-flight CMOS image sensor with pinned-photodiode pixel structure. IEEE Electron Dev. Lett. **31**(11), 1272–1274 (2010)
19. Y. Yan, Q. Chen, L. Wen, H. Xin, H.-F. Zhang, Spatial optical crosstalk in CMOS image sensors integrated with plasmonic color filters. Opt. Express **23**(17), 21994–22003 (2015)
20. M. Miyata, M. Nakajima, T. Hashimoto, High-sensitivity color imaging using pixel-scale color splitters based on dielectric metasurfaces. ACS Photon. **6**(6), 1442–1450 (2019)
21. X. He, N. O'Keefe, Y. Liu, D. Sun, H. Uddin, A. Nirmalathas, R.R. Unnithan, Transmission enhancement in coaxial hole array based plasmonic color filter for image sensor applications. IEEE Photon. J. **10**(4), 1–9 (2018)
22. X. He, N. O'Keefe, D. Sun, Y. Liu, H. Uddin, A. Nirmalathas, R.R. Unnithan, Plasmonic narrow bandpass filters based on metal-dielectric-metal for multispectral imaging, in *CLEO Pacific Rim Conference 2018* (Optical Society of America, 2018), p. Th4E.5
23. R. Rajasekharan, E. Balaur, A. Minovich, S. Collins, T.D. James, A. Djalalian-Assl, K. Ganesan, S. Tomljenovic-Hanic, S. Kandasamy, E. Skafidas, D.N. Neshev, P. Mulvaney, A. Roberts, S. Prawer, Filling schemes at submicron scale: development of submicron sized plasmonic colour filters. Sci. Rep. **4**(1), 6435 (2014)
24. R. Rajasekharan Unnithan, M. Sun, X. He, E. Balaur, A. Minovich, D. N. Neshev, E. Skafidas, A. Roberts, Plasmonic colour filters based on coaxial holes in aluminium. Mater. (Basel) **10**(4) (2017)
25. X. He, Y. Liu, P. Beckett, H. Uddin, A. Nirmalathas, R.R. Unnithan, Transmission enhancement in plasmonic nanohole array for colour imaging applications, vol. 11200, in *AOS Australian Conference on Optical Fibre Technology (ACOFT) and Australian Conference on Optics, Lasers, and Spectroscopy (ACOLS)* eds. by A. Mitchell, H. Rubinsztein-Dunlop (International Society for Optics and Photonics, SPIE, 2019), pp. 227–228

26. B. Zeng, Y. Gao, F.J. Bartoli, Ultrathin nanostructured metals for highly transmissive plasmonic subtractive color filters. Sci. Rep. **3**(1), 2840 (2013)
27. Y. Liang, S. Zhang, X. Cao, L. Yanqing, X. Ting, Free-standing plasmonic metal-dielectric-metal bandpass filter with high transmission efficiency. Sci. Rep. **7**(1), 4357 (2017)
28. A. Maryanto, N. Widijatmiko, W. Sunarmodo, M. Soleh, R. Arief, Development of pushbroom airborne camera system using multispectrum line scan industrial camera. Int. J. Remote Sens. Earth Sci. **13**(1)
29. W.-B. Lee, J. Oliver, S.-C. Kim, H.-N. Lee, Random optical scatter filters for spectrometers: Implementation and estimation, in *Imaging and Applied Optics* (Optical Society of America, 2013), p. JTu4A.33
30. Z. Wang, S. Yi, A. Chen, M. Zhou, T.S. Luk, A. James, J. Nogan, W. Ross, G. Joe, A. Shahsafi, K.X. Wang, M.A. Kats, Z. Yu, Single-shot on-chip spectral sensors based on photonic crystal slabs. Nat. Commun. **10**(1), 1020 (2019)
31. G.R. Arce, D.J. Brady, L. Carin, H. Arguello, D.S. Kittle, Compressive coded aperture spectral imaging: an introduction. IEEE Signal Process. Mag. **31**(1), 105–115 (2014)
32. H. Arguello, G.R. Arce, Rank minimization code aperture design for spectrally selective compressive imaging. IEEE Trans. Image Process. **22**(3), 941–954 (2013)
33. H. Arguello, H. Rueda, W. Yuehao, D.W. Prather, G.R. Arce, Higher-order computational model for coded aperture spectral imaging. Appl. Opt. **52**(10), D12–D21 (2013)
34. H. Arguello, C.V. Correa, G.R. Arce, Fast lapped block reconstructions in compressive spectral imaging. Appl. Opt. **52**(10), D32–D45 (2013)
35. X. Cai, J. Xiong, K. Cui, Y. Huang, H. Zhu, Z. Zheng, S. Xu, Y. He, F. Liu, X. Feng, W. Zhang, One-shot ultraspectral imaging with reconfigurable metasurfaces (2020)
36. W. Zhang, H. Song, X. He, L. Huang, X. Zhang, J. Zheng, W. Shen, X. Hao, X. Liu, Deeply learned broadband encoding stochastic hyperspectral imaging. Light Sci. Appl. **10**(1), 108 (2021)
37. Tetracam Inc. Imaging systems. https://www.tetracam.com/ImagingSystems.htm, 2020. Accessed June 2021
38. MicaSense Inc. Rededge MX. https://micasense.com/rededge-mx/, 2020. Accessed June 2021
39. T. Skauli, An upper-bound metric for characterizing spectral and spatial coregistration errors in spectral imaging. Opt. Express **20**(2), 918–933 (2012)
40. L.L. Coulter, D.A. Stow, Assessment of the spatial co-registration of multitemporal imagery from large format digital cameras in the context of detailed change detection. Sensors **8**(4), 2161–2173 (2008)
41. F. Khaghani, R.J. Nelson, Co-registration of multispectral images for enhanced target recognition, vol. 6565, in *Algorithms and Technologies for Multispectral Hyperspectral, and Ultraspectral Imagery XIII*, eds. by S.S. Shen, P.E. Lewis (International Society for Optics and Photonics, SPIE, 2007), pp. 509–519
42. Y. Han, J. Choi, J. Jung, A. Chang, O. Sungchan, J. Yeom, Automated coregistration of multisensor orthophotos generated from unmanned aerial vehicle platforms. J. Sensors **2019**, 2962734 (2019)
43. IMEC. Hyperspectral imaging. https://www.imec-int.com/en/hyperspectral-imaging, 2021. Accessed June 2021
44. Spectral Devices Inc. Multispectral camera solutions. https://www.spectraldevices.com/, 2021. Accessed June 2021
45. Ocean Insight. Products. https://www.oceaninsight.com/products/, 2021. Accessed June 2021
46. K. Diest, J.A. Dionne, M. Spain, H.A. Atwater, Tunable color filters based on metal-insulator-metal resonators. Nano Lett. **9**(7), 2579–2583 (2009)
47. Y.-J. Jen, A. Lakhtakia, M.-J. Lin, W.-H. Wang, W. Huang-Ming, H.-S. Liao, Metal/dielectric/metal sandwich film for broadband reflection reduction. Sci. Rep. **3**(1), 1672 (2013)
48. V.V. Medvedev, V.M. Gubarev, C.J. Lee, Optical performance of a dielectric-metal-dielectric antireflective absorber structure. J. Opt. Soc. Am. A **35**(8), 1450–1456 (2018)

49. Q. Li, Z. Li, X. Xiang, T. Wang, H. Yang, X. Wang, Y. Gong, J. Gao, Tunable perfect narrow-band absorber based on a metal-dielectric-metal structure. Coatings **9**(6) (2019)
50. J.Y.Y. Loh, N. Kherani, Design of nano-porous multilayer antireflective coatings. Coatings **7**(9) (2017)
51. Y. Matsuoka, S. Mathonnèire, S. Peters, W.T. Masselink, Broadband multilayer anti-reflection coating for mid-infrared range from 7 to 12 μm. Appl. Opt. **7**, 1645–1649
52. Z. Li, Q. Li, X. Quan, X. Zhang, C. Song, H. Yang, X. Wang, J. Gao, Broadband high-reflection dielectric PVD coating with low stress and high adhesion on PMMA. Coatings **9**(4) (2019)
53. C. Hu, J. Liu, J. Wang, Z. Gu, C. Li, Q. Li, Y. Li, S. Zhang, C. Bi, X. Fan, W. Zheng, New design for highly durable infrared-reflective coatings. Light Sci. Appl. **7**
54. S. Pimenta, S. Cardoso, A. Miranda, P. De Beule, E.M.S. Castanheira, G. Minas, Design and fabrication of SiO_2/TiO_2 and Msenmaier, agO/TiO_2 based high selective optical filters for diffuse reflectance and fluorescence signals extraction. Biomed. Opt. Express **6**(8), 3084–3098 (2015)
55. Y.-J. Jen, C.-C. Lee, L. Kun-Han, C.-Y. Jheng, Y.-J. Chen, Fabry-Pérot based metal-dielectric multilayered filters and metamaterials. Opt. Express **23**(26), 33008–33017 (2015)
56. C. Williams, G.S.D. Gordon, T.D. Wilkinson, S.E. Bohndiek, Grayscale-to-color: Scalable fabrication of custom multispectral filter arrays. ACS Photon. **6**(12), 3132–3141 (2019)
57. J.D. Joannopoulos, S.G. Johnson, J.N. Winn, R.D. Meade, *Photonic Crystals: Molding the Flow of Light*, 2nd ed. (Princeton University Press, 2008)
58. S.P. Ojha, Sanjeev K. Srivastava, N. Kumar, S.K. Srivastava, Design of an optical filter using photonic band gap material. Optik **114**(3), 101–105 (2003)
59. Q.Q.X. Changqing, C. Wang, All-dielectric polarization-independent optical angular filter. Sci. Rep. **7**(1), 16574 (2017)
60. M. Scalora, M.J. Bloemer, A.S. Pethel, J.P. Dowling, C.M. Bowden, A.S. Manka, Transparent, metallo-dielectric, one-dimensional, photonic band-gap structures. J. Appl. Phys. **83**(5), 2377–2383 (1998)
61. D.H Cushing, Multilayer thin film dielectric bandpass filter. U.S. Patent US5926317A, 20 Aug 1999 [Online]. Available https://patents.google.com/patent/US6018421A/pt
62. D.H. Cushing, Multilayer thin film bandpass filter. U.S. Patent US6018421A, 25 Jan 2000 [Online]. Available https://patents.google.com/patent/US6018421A/pt
63. R.-Y. Tsai, H.-Y. Lin, Y.-H. Chen, C.-S. Chang, Polarization-independent ultra-narrow band pass filters. U.S. Patent US20020080493A1, 27 June 2003 [Online]. Available https://patents.google.com/patent/US6018421A/pt
64. K.L. Lewis, Multilayer optical filters. U.S. Patent US6631033B1, 7 Oct 2003 [Online]. Available https://patents.google.com/patent/US6631033B1/en?oq=US6631033B1
65. A.C. Kundu, Multilayer band pass filter. U.S. Patent US7312676B2, 25 Dec 2007 [Online]. Available https://patents.google.com/patent/US7312676B2/en?oq=U.S.+Patent+No.+7312676B2
66. S.S. Wang, R. Magnusson, Theory and applications of guided-mode resonance filters. Appl. Opt. **32**(14), 2606–2613 (1993)
67. S. Tibuleac, R. Magnusson, Reflection and transmission guided-mode resonance filters. J. Opt. Soc. Am. A **14**(7), 1617–1626 (1997)
68. A. Sharon, D. Rosenblatt, A.A. Friesem, Narrow spectral bandwidths with grating waveguide structures. Appl. Phys. Lett. **69**(27), 4154–4156 (1996)
69. S. Tibuleac, R. Magnusson, Narrow-linewidth bandpass filters with diffractive thin-film layers. Opt. Lett. **26**(9), 584–586 (2001)
70. S. Babu, J.-B. Lee, Axially-anisotropic hierarchical grating 2d guided-mode resonance strain-sensor. Sensors **19**(23) (2019)
71. W.-K. Kuo, C.-J. Hsu, Two-dimensional grating guided-mode resonance tunable filter. Opt. Express **25**(24), 29642–29649 (2017)
72. Z. Ren, Y. Sun, S. Zhang, K. Zhang, Z. Lin, S. Wang, 2d non-polarising transmission filters based on gmr for optical communications. Micro Nano Lett **13**, 1621–1626(5) (2018)

73. L. Cao, P. Fan, E.S. Barnard, A.M. Brown, M.L. Brongersma, Tuning the color of silicon nanostructures. Nano Lett. **10**(7), 2649–2654 (2010)
74. H. Park, K. Seo, K.B. Crozier, Adding colors to polydimethylsiloxane by embedding vertical silicon nanowires. Appl. Phys. Lett. **101**(19), 193107 (2012)
75. H. Park, K.B. Crozier, Multispectral imaging with vertical silicon nanowires. Sci. Rep. **3**(1), 2460 (2013)
76. H. Park, Y. Dan, K. Seo, Y.J. Yu, P.K. Duane, M. Wober, K.B. Crozier, Filter-free image sensor pixels comprising silicon nanowires with selective color absorption. Nano Lett. **14**(4), 1804–1809 (2014)
77. N. Dhindsa, J. Walia, M. Pathirane, I. Khodadad, W.S. Wong, S.S. Saini, Adjustable optical response of amorphous silicon nanowires integrated with thin films. Nanotechnology **27**(14), 145703 (2016)
78. H.S. Song, G.J. Lee, D.E. Yoo, Y.J. Kim, Y.J. Yoo, D.-W. Lee, V. Siva, Il-Suk Kang, and Young Min Song. Reflective color filter with precise control of the color coordinate achieved by stacking silicon nanowire arrays onto ultrathin optical coatings. *Scientific Reports*, 9(1):3350, 2019
79. H.A. Bethe, Theory of diffraction by small holes. Phys. Rev. **66**, 163–182 (1944)
80. T.W. Ebbesen, H.J. Lezec, H.F. Ghaemi, T. Thio, P.A. Wolff, Extraordinary optical transmission through sub-wavelength hole arrays. Nature **391**(6668), 667–669 (1998)
81. S.A. Maier, *Plasmonics—Fundamentals and Applications*, 1st. edn. (Springer, 2007)
82. J.A. Porto, F.J. Garcá-Vidal, J.B. Pendry, Transmission resonances on metallic gratings with very narrow slits. Phys. Rev. Lett **83**, 2845 (1999)
83. T. Xu, Y. Wu, X. Luo, L.J. Guo, Plasmonic nanoresonators for high-resolution color filtering and spectral imaging. Nat Commun **1**, 59 (2010)
84. H.-S. Lee, Y.-T. Yoon, S.-S. Lee, S.-H. Kim, K.-D. Lee, Color filter based on a subwavelength patterned metal grating. Opt. Express **15**(23), 15457–15463 (2007)
85. D. Fleischman, L.A. Sweatlock, H. Murakami, H. Atwater, Hyper-selective plasmonic color filters. Opt. Express **25**, 27386–27395 (2017)
86. D. Fleischman, K.T. Fountaine, C.R. Bukowsky, G. Tagliabue, L.A. Sweatlock, H.A. Atwater, High spectral resolution plasmonic color filters with subwavelength dimensions. ACS Photon. **6**(2), 332–338 (2019)
87. D.B. Mazulquim, K.J. Lee, J.W. Yoon, L.V. Muniz, B.-H.V. Borges, L.G. Neto, R. Magnusson, Efficient band-pass color filters enabled by resonant modes and plasmons near the rayleigh anomaly. Opt. Express **22**, 30843–30851 (2014)
88. H.D. Lang, B. Gijsbertus, Optical system for a color television camera, Aug 1961
89. P.-J. Lapray, X. Wang, J.-B. Thomas, P. Gouton, Multispectral filter arrays: recent advances and practical implementation. Sensors **14**(11), 21626–21659 (2014)
90. S.P. Burgos, S. Yokogawa, H.A. Atwater, Color imaging via nearest neighbor hole coupling in plasmonic color filters integrated onto a complementary metal-oxide semiconductor image sensor. ACS Nano **7**(11), 10038–10047 (2013)
91. M. Najiminaini, F. Vasefi, B. Kaminska, J.J.L. Carson, Nanohole-array-based device for 2D snapshot multispectral imaging. Sci. Rep. **3**, 2589 (2013)
92. W. Jang, Z. Ku, J. Jeon, J.O. Kim, S.J. Lee, J. Park, M.J. Noyola, A. Urbas, Experimental demonstration of adaptive infrared multispectral imaging using plasmonic filter array. Sci. Rep. **6**, 34876 (2016)
93. A. Manjavacas, F.J. Garcia de Abajo, Tunable plasmons in atomically thin gold nanodisks. Nat. Commun. **5**, 3548 (2014)
94. C. Langhammer, M. Schwind, B. Kasemo, I. Zori, Localized surface plasmon resonances in aluminum nanodisks. Nano Lett. **8**(5), 1461–1471 (2008)
95. J.S. Clausen, E. Hijlund-Nielsen, A.B. Christiansen, S. Yazdi, M. Grajower, H. Taha, U. Levy, A. Kristensen, N.A. Mortensen, Plasmonic metasurfaces for coloration of plastic consumer products. Nano Lett. **14**(8), 4499–4504 (2014)
96. M. Ye, L. Sun, X. Hu, B. Shi, B. Zeng, L. Wang, J. Zhao, S. Yang, R. Tai, H.-J. Fecht, J.-Z. Jiang, D.-X. Zhang, Angle-insensitive plasmonic color filters with randomly distributed silver nanodisks. Opt. Lett. **40**, 4979–4982 (2015)

97. K.C. Chua, C.Y. Chao, Y.F. Chen, Electrically controlled surface plasmon resonance frequency of gold nanorods. Appl. Phys. Lett **89**(7), 10310 (2006)

98. J. Zhang, M. ElKabbash, R. Wei, S. C. Singh, B. Lam, C. Guo, Plasmonic metasurfaces with 42.3% transmission efficiency in the visible. Light Sci. Appl. **8**, 53 (2019)

99. B. Heshmat, D. Li, T.E. Darcie, R. Gordon, Tuning plasmonic resonances of an annular aperture in metal plate. Opt. Express **19**, 5912–5923 (2011)

100. J. Salvi, M. Roussey, F.I. Baida, M.-P. Bernal, A. Mussot, T. Sylvestre, H. Maillotte, D. Van Labeke, A. Perentes, I. Utke, C. Sandu, P. Hoffmann, B. Dwir, Annular aperture arrays: study in the visible region of the electromagnetic spectrum. Opt. Lett **30**, 1611–1613 (2005)

101. G. Si, Y. Zhao, H. Liu, S. Teo, M. Zhang, T.J. Huang, A.J. Danner, J. Teng, Annular aperture array based color filter. Appl. Phys. Lett **99**, 033105 (2011)

102. X. He, Y. Liu, K. Ganesan, A. Ahnood, P. Beckett, F. Eftekhari, D. Smith, M.H. Uddin, E. Skafidas, A. Nirmalathas, R.R. Unnithan, A single sensor based multispectral imaging camera using a narrow spectral band color mosaic integrated on the monochrome CMOS image sensor. APL Photon **5**(4), 046104 (2020)

103. S. Yokogawa, S.P. Burgos, H.A. Atwater, Plasmonic color filters for CMOS image sensor applications. Nano Lett. **12**(8), 4349–4354 (2012)

104. Q. Chen, D.R.S. Cumming, High transmission and low color cross-talk plasmonic color filters using triangular-lattice hole arrays in aluminum films. Opt. Express **18**, 14056–14062 (2010)

105. L. Duempelmann, B. Gallinet, L. Novotny, Multispectral imaging with tunable plasmonic filters. ACS Photon. **4**(2), 236–241 (2017)

Chapter 3
Transmission Enhancement in Coaxial Hole Array Based Plasmonic Color Filters

Plasmonic color filters based on the hexagonal arrangement of coaxial hole array in aluminium have been demonstrated to be a promising candidate for color filter development due to their polarization and incident angle insensitivity. They comprise a single nanometer thick metal film, and demonstrate good line width, CMOS compatibility and fine color tunability. However, the low transmittance of the coaxial hole array based filters limits their use in potential applications such as image sensors and displays. This chapter introduces a new method to increase transmittance in coaxial hole array based filters by tuning both localized surface plasmons and surface plasmon polaritons. In this method, small diameter coaxial holes are created along with a large size coaxial hole array to form a combined filter geometry and computational techniques are employed to optimize the parameters within the aluminium film. The simulation results will have potential applications in CMOS image sensors, submicron pixel development, micro-displays and liquid crystal over silicon (LCoS) devices.

3.1 Introduction

As demand for high-resolution imagery continues to expand in consumer, scientific and industrial applications there is a need for smaller, highly tunable and robust color filters [1–4]. Precise control of wavelength provides vivid color response and improved selectivity. CMOS image sensor technologies have undergone enormous advances in recent years to increase performance and reduce pixel size. Along with reducing the transistor size to nanoscale for image sensors, submicron scale filters are also required to perform filtering for image acquisition sensors, next generation spatial light modulators [5–7] and high-resolution micro-displays.

© The Author(s), under exclusive license to Springer Nature Singapore Pte Ltd. 2021
X. He et al., *Multispectral Image Sensors Using Metasurfaces*, Progress in Optical Science and Photonics 17, https://doi.org/10.1007/978-981-16-7515-7_3

As conventional color filters are made from pigments and dyes, their color response is directly related to the absorption coefficient, which in turn is related to the area/volume of the filter materials. Therefore, it is difficult to make nanometer-sized color filters using pigments and dyes [1–4, 8, 9]. Recently, color filters based on TiO$_2$ nanodisks operating in reflection mode were reported as having good optical performance because of their low loss and high refractive index in the visible range [4]. However, this design is not suitable for image sensor applications due to their operation in reflection mode and the TiO$_2$ nanodisk material is not CMOS compatible.

Plasmonic color filters based on aluminium are better candidates for use in most of the all-dielectric color filters due to their CMOS compatibility, ability to operate in transmission mode, use of single namometer thick film and excellent color tunability (from UV to NIR range) [1]. Furthermore, the filter spectral response can be carefully controlled by parameterizing the geometry, producing artificial material properties engineered to specific design and application requirements. For this reason, there have been many demonstrations of color filters employing plasmonic effects, particularly using nanostructures in thin metal films [1, 2, 4, 8–28]. Color filters such as these can be fabricated using a single perforated metal film with the thickness of 50 to 200 nm.

Despite these promising attributes, plasmonic filters still suffer from poor transmission and sensitivity to angle of incidence which have limited the scope of target applications and increased deployment requirements in CMOS and display applications. For example, the peak transmittance wavelength for a simple hole array based plasmonic color filters depends on the angle of the incident light because the surface plasmon based effect in the hole array is angle sensitive [8, 29]. If such a filter were used in an image sensor, the filter will transmit only the desired color at normal angles of incidence while different colors would be filtered at oblique angles.

Plasmonic color filters based on an array of coaxial holes have been reported using gold (Au) [30]. However, the coaxial hole array in an Au film is limited to produce a resonance peak at wavelengths less than 480 nm due to material limitations [31]. It has been demonstrated that the transmittance of a square aperture array in silver film can be enhanced using suitably placed small squares [32]. However, the square aperture is polarization sensitive. Other designs were reported that create symmetrical patterns inserted into a metal film to increase the transmittance, such as nano-trench or cavity structures [33, 34]. An extra coating on the Ag–Si film was also demonstrated to enhance the transmittance [35] based on Fabry-Pérot resonances. However, these demonstrated structures do not cover the full gamut of the visible range, and Ag quickly oxidizes in air, degrading its optical characteristics. Further, Au and Ag based materials are not CMOS compatible and require a seed layer for better adhesion to silicon substrates, which increases the fabrication complexity.

Recently, it has been reported that vivid red, green and blue colors were able to be produced by plasmonic color filters based on the hexagonal arrangement of coaxial holes in aluminium [16]. This is a promising candidate for image sensor applications as its characteristics are insensitive to the angle of incidence and independent of the polarization direction. However, the transmittance of color filters based on coaxial hole arrays is too low to be used in an image sensor or display as the transmittance reported in the literature varies from 5% (red) to little more than 20% (blue) [16]. This is because the peak wavelength of the localized surface plasmon depends on the gap size and so, as the gap decreases, a shift towards the red region is introduced. Therefore, it is important to develop techniques to increase the transmittance of plasmonic color filters without compromising important characteristics such as polarization and incident angle insensitivity, good line width, fine tuning capability and CMOS compatibility.

This chapter presents a technique to enhance the transmittance of plasmonic color filters based on the coaxial hole array in aluminium to overcome the limitation of low transmittance in the coaxial hole array filter using two different coaxial holes geometry that are independently tuned to the same color (wavelength), followed by merging both the coaxial geometries to make a combined geometry.

3.2   Simulation Model and Optimization Method

The plasmonic resonance peaks responsible for obtaining different colors in a coaxial geometry are predominantly due to localized surface plasmons (Fabry-Pérot resonances) supported in a cylindrical resonance cavity. The cavity is excited by cylindrical surface plasmons formed by a metal film with finite thickness and two end faces. Tuning of colors was obtained by varying the inner (r_1) and outer (r_2) radii of the coaxial hole using localized surface plasmon resonance. The resonance peak (peak transmittance wavelength) can be obtained using different combinations of the inner (r_1) and outer radii (r_2) as shown in Fig. 3.1. The desired color (transmittance peak) can be calculated using: $l = (n\pi - \Omega)/\beta$ [8, 9, 36], where l, n, Ω and β are the thickness of the metal film, the order of the Fabry-Pérot resonance, the phase of reflection constant and the propagation constant, respectively. There is also a minor contribution to resonance peaks in the coaxial hole array geometry due to the pitch between coaxial holes caused by surface plasmons polaritons [16, 30]. The coaxial hole array in the hexagonal arrangement was simulated using a unit cell of a regular hexagon as shown in Fig. 3.1a where W and L are defined as the separation between large coaxial holes in x and y directions respectively. The relationship between W and L are chosen such that:

$$L = 2 \times W \times \frac{\sqrt{3}}{2} = \sqrt{3}W. \qquad (3.1)$$

To increase the transmittance, the hexagonal geometry of the large coaxial holes array (LCHA in the blue regular hexagon block in Fig. 3.1a) is combined with an additional coaxial hole array with a smaller radius (SCHA in Fig. 3.1b) to create a dual coaxial hole array (DCHA) in such a way that the combined geometry is polarization insensitive. In the DCHA, the distance between two adjacent small coaxial holes is

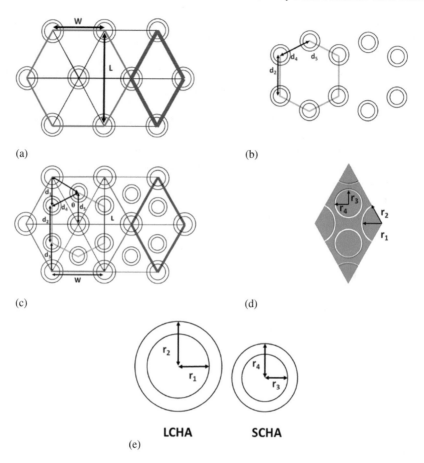

Fig. 3.1 Schematic of **a** large coaxial hole array (LCHA): coaxial hole array showing a unit cell (red) used for simulation. Periodic boundary condition was used on four sides to simulate an array, **b** small coaxial hole array (SCHA), **c** dual coaxial hole array (DCHA) made of a combination of large coaxial hole array (LCHA) and small coaxial hole array (SCHA), **d** top view of DCHA unit cell, **e** the radius of LCHA and SCHA

the same as the length d_1, the distance between the adjacent large coaxial hole and small coaxial hole. With the angle $\theta = 60°$, we get:

$$d_1 = d_2 = d_3 = d_4 = d_5 = \frac{1}{3}L = \frac{\sqrt{3}}{3}W. \tag{3.2}$$

These 3-D geometries were computationally investigated using finite element methods (FEM) implemented in COMSOL MULTIPHYSICS®.[1] The simulation unit cell is identified by the red diamond block in Fig. 3.1c and its top view is shown

---

[1] See Appendix A for more details of the COMSOL FEM simulation methods.

Fig. 3.2 Schematic of
incident plane wave: k is the
propagation constant, θ is the
angle of incidence and ϕ is
the polarization angle

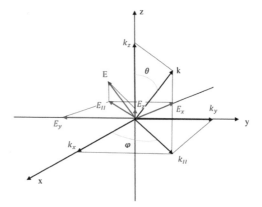

in Fig. 3.1d. The simulation model comprises a 150 nm thick layer of aluminium
on a semi-infinite glass substrate (n = 1.5) 200 nm thick. The top layer is air with
a thickness of 200 nm. Light is excited from the aluminium side (top side) using
port boundary conditions. The overall objective is to find the wavelength at which
maximum transmittance occurs for the filters and so S-parameters are used to find the
transmittance ($|S21|^2$) of the filters. Perfectly matched layer (PML) is used on the top
and at the bottom of the model to avoid effects of reflected light in the transmittance
spectrum.

Details of the incident wave are shown in Fig. 3.2, and the electric field equations
used to study polarization and angle sensitivity are as follows [37]:

$$E_x = A \sin\left(\frac{\pi}{2} - \theta\right) \cos(\pi + \varphi) e^{-j(k_x x + k_y y + K_z z)}$$

$$E_y = A \sin\left(\frac{\pi}{2} - \theta\right) \sin(\pi + \varphi) e^{-j(k_x x + k_y y + K_z z)}$$

$$E_z = A \cos\left(\frac{\pi}{2} - \theta\right) e^{-j(k_x x + k_y y + K_z z)} \tag{3.3}$$

$$|E| = \sqrt{E_x^2 + E_y^2 + E_z^2} = A,$$

$$k_x = k \sin\theta \cos\varphi, \, k_y = k \sin\theta \sin\varphi, \, k_z = k \cos\theta.$$

where θ is the angle of incidence, and ϕ is the polarization angle which is in the direction of E. The propagation constant k is defined as $(2\pi n)/\lambda$ where n is the refractive
index of the simulation domain (air) where the port is set for illumination. The refractive indices of aluminium at different wavelengths are taken from Palik's experiment
[38]. The angle of incidence θ was varied to study its effect of transmittance and,
similarly, the polarization angle ϕ was varied to study the polarization sensitivity.

3.3 Experimental Results

Based on the simulation models discussed above, in the first step, a large coaxial hole array was designed to produce a peak at 650nm (red color filter) by tuning the inner (r_1) and outer (r_2) radii to 120nm and 130nm, respectively by tuning the localized surface plasmons. This is followed by adjusting the pitch to achieve 420nm for $LCHA_R$ by tuning the LSP. $LCHA_R$ gives a transmittance of 6.2% and a line width of 180nm (FWHM). In the second step, another small coaxial hole array is designed to the same peak wavelength of 650nm by tuning inner and outer radii to 83nm and 90nm, respectively ($SCHA_R$). $SCHA_R$ gives a 5.6% transmittance with FWHM of 150nm. In the final step, both $LCHA_R$ and $SCHA_R$ are combined to create the dual coaxial hole array ($DCHA_R$) with the enhanced transmittance of 15.5% with FWHM of 215nm. These results are presented in Fig. 3.3. The insets of Fig. 3.3 show the normalized electric field on $LCHA_R$, $SCHA_R$ and $DCHA_R$ which shows that both large and small coaxial hole arrays are excited by the incident light, and both are contributing to transmission. The increased transmittance is due to the coupling of plasmon resonances in both large and small coaxial holes constructively by geometrical tuning.

RGB colors are required for making the Bayer pattern in CMOS image sensors. Hence the same scheme used for red is applied for both green and blue color filters. The blue color ($LCHA_B$) is obtained by tuning the pitch size to 260nm, and radii to $r_1 = 40$nm and $r_2 = 50$nm. The $LCHA_B$ is then combined with a small coaxial hole array tuned to the same wavelength ($SCHA_B$) with $r_3 = 40$nm, $r_4 = 50$nm to obtain

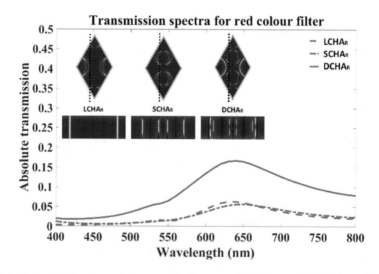

Fig. 3.3 Red color filter: transmission spectra of large coaxial hole array ($LCHA_R$), small coaxial hole array ($SCHA_R$) and dual coaxial hole array ($DCHA_R$). The inset images show the normalized electric field from top view and lateral cross section along the dotted black line for $LCHA_R$, $SCHA_R$ and $DCHA_R$.

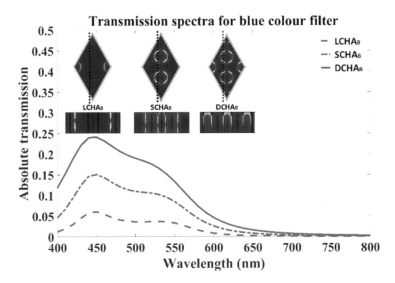

Fig. 3.4 Blue color filter: transmission spectra of large coaxial hole array (LCHA$_B$), small coaxial hole array (SCHA$_B$) and dual coaxial hole array (DCHA$_B$). The inset images show the normalized electric field from top view and lateral cross section along the dotted black line for LCHA$_B$, SCHA$_B$ and DCHA$_B$

DCHA$_B$. For the green LCHA$_G$, the pitch size is 430 nm, r_1 and r_2 are 106 nm and 130 nm respectively, which is then combined with another smaller coaxial hole array (SCHA$_G$) with $r_3 = 70$ nm, $r_4 = 80$ nm to create the green color array (DCHA$_G$).

The transmission spectra of the blue and green arrays (LCHA, SCHA and DCHA) are shown in Figs. 3.4 and 3.5 respectively, and their line widths (FWHM) are tabulated in Table 3.1. A CIE 1931 chromaticity diagram was plotted for the red, green and blue color filters using chromaticity coordinates obtained from simulation results for the LCHA and DCHA (Fig. 3.6). The CIE chromaticity chart shows that the pure color perceived by human vision is obtained from the spectrum of different color filters. From the CIE chromaticity charts of LCHA and DCHA, it can be seen that the measured chromaticity coordinates of the RGB filters falls around the achromatic point. For the green and red DCHA color filters, their positions are slightly shifted on the CIE chart towards the center (white light) compared to SCHA. The shift is due to a slight increase in FWHM after combining the LCHA and SCHA, but the positions still fall near the achromatic point and are within acceptable limits.

This demonstrates the capability of DCHA based color filters with increased transmittance for a large degree of color range tuning applications such as CMOS image sensors.

There are two important requirements for a color filter to be used as a CMOS image sensor: polarization insensitivity and angle independence. The polarization insensitivity ensures that the exact color and intensity of objects are captured irrespective of the rotation of camera/sensor. Angle independence is essential to obtain

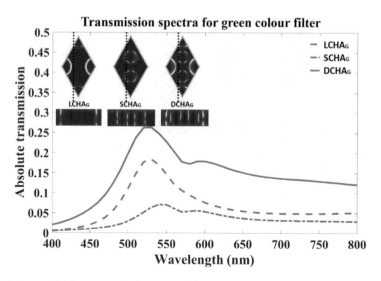

Fig. 3.5 Green color filter: transmission spectra of large coaxial hole array (LCHA$_G$), small coaxial hole array (SCHA$_G$) and dual coaxial hole array (DCHA$_G$). The inset images show the normalized electric field from top view and lateral cross section along the dotted black line for LCHA$_G$, SCHA$_G$ and DCHA$_G$

Table 3.1 Maximum transmittance, peak wavelength and FWHM for LCHA, SCHA and DCHA of blue, green and red color filters

Color	Model	Maximum-transmittance (%)	Peak wavelength (nm)	FWHM (nm)
Blue	LCHA$_B$	6.0	450	140
	SCHA$_B$	15.0	450	150
	DCHA$_B$	24	450	150
Green	LCHA$_G$	18.4	530	120
	SCHA$_G$	7.1	530	180
	DCHA$_G$	26.4	530	200
Red	LCHA$_R$	6.2	650	180
	SCHA$_R$	5.6	650	150
	DCHA$_R$	15.5	650	215

the same color of objects irrespective of the angle of recording. For example, if the filters are angle sensitive, the same object will be recorded with different colors at different angles of the image capture. The transmittance of the DCHA was studied for three different polarization angles, 0°, 45° and 90°. A red DCHA filter (DCHA$_R$) has been taken as an example and shown in Fig. 3.7a. The results illustrate that the DCHA is almost polarization insensitive. Furthermore, transmission characteristics of the red DCHA were studied for different incident angles (full field angle: FFOV), 0°, 30° and 60°, as shown in Fig. 3.7b. There is no shift in the peak wavelength with

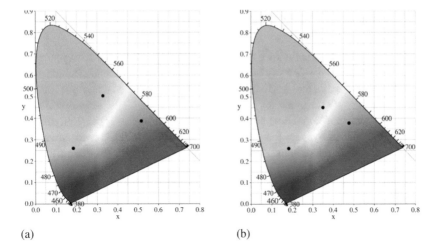

Fig. 3.6 CIE chromaticity chart of blue, green and red color filter for **a** LCHA, and **b** DCHA

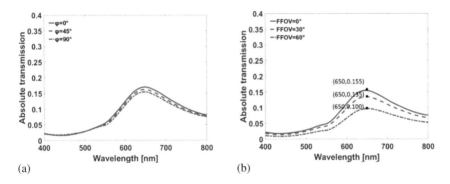

Fig. 3.7 Transmission spectrum of DCHA for different polarizations and angle of excitations **a** the polarization angles $\phi = 0°$, $\phi = 45°$, $\phi = 90°$ and corresponding transmission spectra showing DCHA is polarization insensitive, **b** angle dependence of DCHA shows no shift in peak wavelength with respect to angle and variation in transmittance is negligible. Here, the angle of incidence (excitation) represented as full field of view (full angle), FFOV = 0°, 30°, 60°

respect to angle, and hence the transmitted color will be the same from different angles of illumination with a small reduction in the intensity ($\approx 3.5\%$). These results show that the DCHA design is very promising for color filter development due to its insensitivity to incidence angle and polarization, good line width (FWHM), fine tuning ability and CMOS compatibility.

3.4   Summary

The chapter presents a technique to increase peak transmittance of the hexagonal arrangement of coaxial hole array in aluminum without compromising its polarization insensitivity, line width and angle insensitivity. In the proposed technique, one large coaxial hole array is designed for a peak wavelength of interest after optimization, followed by a small coaxial hole array designed to the same peak wavelength. These two coaxial hole arrays are then combined in a suitable fashion to enhance the transmittance. The transmittance is increased more than $2\times$ for red color filter, $1.4\times$ for green and $4\times$ for blue. These results will have potential applications in the development of micron and submicron scale color filter arrays for CMOS sensors, micro-displays and spatial light modulators.

References

1. S. Yokogawa, S.P. Burgos, H.A. Atwater, Plasmonic color filters for CMOS image sensor applications. Nano Lett. **12**(8), 4349–4354 (2012)
2. S.P. Burgos, S. Yokogawa, H.A. Atwater, Color imaging via nearest neighbor hole coupling in plasmonic color filters integrated onto a complementary metal-oxide semiconductor image sensor. ACS Nano **7**(11), 10038–10047 (2013)
3. K. Diest, J.A. Dionne, M. Spain, H.A. Atwater, Tunable color filters based on metal-insulator-metal resonators. Nano Lett. **9**(7), 2579–2583 (2009)
4. Y. Yan, Q. Chen, L. Wen, H. Xin, H.-F. Zhang, Spatial optical crosstalk in CMOS image sensors integrated with plasmonic color filters. Opt. Express **23**(17), 21994–22003 (2015)
5. R. Rajasekharan, Q. Dai, T.D. Wilkinson, Electro-optic characteristics of a transparent nanophotonic device based on carbon nanotubes and liquid crystals. Appl. Opt. **49**(11), 2099–2104 (2010)
6. R. Rajasekharan, T.D. Wilkinson, P.J.W. Hands, Q. Dai, Nanophotonic three-dimensional microscope. Nano Lett. **11**(7), 2770–2773 (2011)
7. Q. Dai, R. Rajasekharan, H. Butt, K. Won, X. Wang, T.D. Wilkinson, G. Amaragtunga, Transparent liquid-crystal-based microlens array using vertically aligned carbon nanofiber electrodes on quartz substrates. Nanotechnology **22**(11), 115201 (2011)
8. Q. Chen, D. Chitnis, K. Walls, T.D. Drysdale, S. Collins, D.R.S. Cumming, CMOS photodetectors integrated with plasmonic color filters. IEEE Photon. Technol. Lett. **24**(3), 197–199 (2012)
9. B.Y. Zheng, Y. Wang, P. Nordlander, N.J. Halas, Color-selective and CMOS-compatible photodetection based on aluminum plasmonics. Adv. Mater **26**, 6318–6323 (2014)
10. D. Inoue, A. Miura, T. Nomura, H. Fujikawa, K. Sato, N. Ikeda, D. Tsuya, Y. Sugimoto, Y. Koide, Polarization independent visible color filter comprising an aluminum film with surface-plasmon enhanced transmission through a subwavelength array of holes. Appl. Phys. Lett. **98**(093113) (2011)
11. C. Genet, T. Ebbesen, Light in tiny holes. Nature **445**, 39–46 (2007)
12. L.B. Sun, X.L. Hu, B.B. Zeng, L.S. Wang, S.M. Yang, R.Z. Tai, H.J. Fecht, D.X. Zhang, J.Z. Jiang, Effect of relative nanohole position on colour purity of ultrathin plasmonic substractive colour filters. Nanotechnology **26**(30) (2015)
13. G. Si, Y. Zhao, H. Liu, S. Teo, M. Zhang, T.J. Huang, A.J. Danner, J. Teng, Annular aperture array based color filter. Appl. Phys. Lett **99**, 033105 (2011)

14. T.H. Hsu, Y.K. Fang, C.Y. Lin, S.F. Chen, C.S. Lin, D.N. Yaung, S.G. Wuu, H.C. Chien, C.H. Tseng, J.S. Lin, C.S. Wang, Light guide for pixel crosstalk improvement in deep submicron CMOS image sensor. IEEE Electron Dev. Lett. **25**(1), 22–24 (2004)
15. R. Rajasekharan, E. Balaur, A. Minovich, S. Collins, T.D. James, A. Djalalian-Assl, K. Ganesan, S. Tomljenovic-Hanic, S. Kandasamy, E. Skafidas, D.N. Neshev, P. Mulvaney, A. Roberts, S. Prawer, Filling schemes at submicron scale: development of submicron sized plasmonic colour filters. Scientific Reports **4**(1), 6435 (2014)
16. R.R. Unnithan, M. Sun, X. He, E. Balaur, A. Minovich, D. N. Neshev, E. Skafidas, A. Roberts. Plasmonic colour filters based on coaxial holes in aluminium. Mater. (Basel) **10**(4) (2017)
17. W. Liu, A. Sukhorukov, A. Miroshnichenko, C. Poulton, Z. Y. Xu, D. Neshev, and Y. Kivshar. Complete spectral gap in coupled dielectric waveguides embedded into metal. *Appl. Phys. Lett*, 97(021106), 2010
18. Q. Chen, D.R.S. Cumming, High transmission and low color cross-talk plasmonic color filters using triangular-lattice hole arrays in aluminum films. Opt. Express **18**, 14056–14062 (2010)
19. M. Najiminaini, F. Vasefi, B. Kaminska, J.J.L. Carson, Nanohole-array-based device for 2D snapshot multispectral imaging. Sci Rep **3**, 2589 (2013)
20. H.-S. Lee, Y.-T. Yoon, S.-S. Lee, Sang-Hoon. Kim, Ki-Dong. Lee, Color filter based on a subwavelength patterned metal grating. Opt. Express **15**(23), 15457–15463 (Nov 2007)
21. H.J. Lezec, T. Thio, Diffracted evanescent wave model for enhanced and suppressed optical transmission through subwavelength hole arrays. Opt. Express. **12**(16), 3629–3651 (2004)
22. S. Balakrishnan, M. Najiminaini, M.R. Singh, A.J.J.L. Carson, study of angle dependent surface plasmon polaritons in nano-hole array structures. J. Appl. Phys. **120**, 034302 (2016)
23. B. Zeng, Y. Gao, F.J. Bartoli, Ultrathin nanostructured metals for highly transmissive plasmonic subtractive color filters. Sci. Rep. **3**(1), 2840 (2013)
24. X.N. Zhang, G.Q. Liu, Z.Q. Liu, Z.J. Cai, Y. Hu, X.S. Liu, G.L. Fu, H.G. Gao, S. Huang, Effects of compound rectangular subwavelength hole arrays on enhancing optical transmission. IEEE Photon. J. **7**(1), 1–8 (2015)
25. C.M. Wang, H.I. Huang, C.C. Chao, J.Y. Chang, Y. Sheng, Transmission enhancement through a trench-surrounded nano metallic slit by bump reflectors. Opt. Express **15**, 6 (2007)
26. H. Lu, X.M. Liu, Y.K. Gong, D. Mao, L.R. Wang, Enhancement of transmission efficiency of nanoplasmonic wavelength demultiplexer based on channel drop filters and reflection nanocavities. Opt. Express **19**(14), 12885–90 (2011)
27. Q. Li, K.K. Du, K.N. Mao, X. Fang, D. Zhao, H. Ye, M. Qiu, Transmission enhancement based on strong interference in metal-semiconductor layered film for energy harvesting. Sci. Rep. **6**, 29195 (2016)
28. B. Heshmat, D. Li, T.E. Darcie, R. Gordon, Tuning plasmonic resonances of an annular aperture in metal plate. Opt. Express **19**, 5912–5923 (2011)
29. C. Yang, W. Shen, Y. Zhang, H. Peng, X. Zhang, X. Liu, Design and simulation of omni-directional reflective color filters based on metal-dielectric-metal structure. Opt. Express **22**, 11384–11391 (2014)
30. Astrosurf. RGB versus CMY color imagery. http://www.astrosurf.com/buil/us/cmy/cmy.htm. Accessed Jan 2021
31. T.W. Ebbesen, H.J. Lezec, H.F. Ghaemi, T. Thio, P.A. Wolff, Extraordinary optical transmission through sub-wavelength hole arrays. Nature **391**(6668), 667–669 (1998)
32. H. Wang, X. Wang, C. Yan, H. Zhao, J. Zhang, C. Santschi, O.J.F. Martin, Full color generation using silver tandem nanodisks. ACS Nano. **11**(5), 4419–4427 (2017)
33. M.W. Knight, N.S. King, L. Liu, H.O. Everitt, P. Nordlander, N.J. Halas, Aluminum for plasmonics. ACS Nano. **8**(1), 834–840 (2014)
34. Y.-J. Jen, A. Lakhtakia, M.-J. Lin, W.-H. Wang, W. Huang-Ming, H.-S. Liao, Metal/dielectric/metal sandwich film for broadband reflection reduction. Sci. Rep. **3**(1), 1672 (2013)
35. V.V. Medvedev, V.M. Gubarev, C.J. Lee, Optical performance of a dielectric-metal-dielectric antireflective absorber structure. J. Opt. Soc. Am. A **35**(8), 1450–1456 (2018)
36. H.A. Bethe, Theory of diffraction by small holes. Phys. Rev. **66**, 163–182 (1944)

37. Q. Li, Z. Li, X. Xiang, T. Wang, H. Yang, X. Wang, Y. Gong, J. Gao, Tunable perfect narrow-band absorber based on a metal-dielectric-metal structure. Coatings **9**(6) (2019)
38. T. Pakizeh, A. Dmitriev, M.S. Abrishamian, N. Granpayeh, M. Käll, Structural asymmetry and induced optical magnetism in plasmonic nanosandwiches. J. Opt Soc. Am. B **25**(4), 659–667 (2008)

Chapter 4
CMY Camera Using Nanorod Filter Mosaic Integrated on a CMOS Image Sensor

A CMY color camera differs from its RGB counterpart in that it uses the subtractive colors cyan, magenta and yellow. CMY cameras can perform better than RGB in low light conditions. However, conventional CMY color filter technology made of pigments and dyes are limited in performance for the next generation image sensors with submicron pixel sizes. This is because the conventional CMY filters cannot be fabricated in nanoscale as they use their absorption properties to subtract colors. This chapter presents a CMY camera developed using nanoscale CMY color filter mosaic size of 4 mm × 4 mm with with 4.4 μm color filter blocks made of Al–TiO$_2$–Al nanorods integrated on a commercial CMOS image sensor. Color imaging is demonstrated using a 12 color Macbeth chart. The technology thus developed will have applications in astronomy, low exposure time imaging in biology and photography.[1]

4.1 Introduction

In a conventional CMOS based image sensor, color imaging relies on the integration of filters on top of the photodetector array. These filters typically cover the three primary colors (bands), red (around 650 nm), green (550 nm) and blue (450 nm) (RGB), predominately in a Bayer pattern [2–5, 5–12]. As the human eye is more sensitive to green light than either red or blue, the widely used Bayer filter mosaic is formed with twice as many green as red or blue filters. RGB color space uses an additive color mixing of red, green and blue that combine to create a white output.

---

[1] Material in this chapter from [1], ©The Optical Society (United States), reproduced with permission.

X. He et al., *Multispectral Image Sensors Using Metasurfaces*, Progress in Optical Science and Photonics 17, https://doi.org/10.1007/978-981-16-7515-7_4

In contrast, CMY (cyan, magenta and yellow) is a subtractive color mixing scheme where color filters are used to remove certain wavelengths of white light. For example, cyan is obtained when the red is subtracted from an image. Similarly, magenta and yellow are obtained by subtracting green and blue respectively. In the RGB color space, a red filter transmits only about 1/3 of the visible light as the remaining light is absorbed in the filter (i.e., only the red light passes through the filter). In contrast, the corresponding cyan filter transmits about 2/3 of the spectrum because only the red is subtracted and the remaining is transmitted through the filter. On average, CMY color filters pass approximately twice the spectral power as the corresponding RGB filters [13] and so are promising candidates for low-light imaging applications. Examples include astronomical imaging, of nebula and the like, and are characterized by weak images against a dark background. Further, many astronomical objects need to be captured with short exposure times, before rotation effects blur the image (e.g. photographing planets with rapid rotation such as Jupiter). Finally, images taken with sufficiently short exposure times can avoid the effects of turbulence etc. in the Earth's atmosphere [14]. All of these considerations lead to the need for high transmission filter schemes such as CMY.

Conventional CMY color filters are made of organic dye-based materials or pigments that use their absorption properties to subtract specific wavelengths. However, as the pixel size in the image sensor is reduced to a submicron dimensions, conventional CMY color filters start to suffer from color crosstalk as they are not able to operate at nanoscale thickness [2–5, 5–10]. Further, existing CMY technology must be fabricated in several steps, which presents severe challenges when trying to accomplish submicron scale alignment.

Color filters based on plasmonic effects [2–5, 5–13, 15–27] are suitable for CMY color filter development due to their precise color tuning using nanoscale thick films. Color filters based on localized surface plasmons are superior to filters based on surface plasmons as the former is angle independent. The angle independence of LSP based filters ensures that the transmitted colors are the same from any angle of excitation and this is an important requirement for image sensor applications. Recently, gold (Au) nano-disk based CMY filters have been demonstrated with good characteristics such as polarization independence and angle insensitivity [18]. However, Au based plasmonic color filters exhibit limited color tuning capabilities, especially below 550 nm, as Au does not readily support plasmonic resonance peaks below this wavelength [28]. CMY filters based on silver (Ag) nanoslit [29] and nano-disk [20] have been reported with high transmission coefficient, but Ag oxidizes quickly in air which degrades its optical properties. It therefore requires an additional coating that can affect the filter characteristics. CMY color filters operating in reflection mode based on surface plasmon polaritons (SPP) are demonstrated in [22–25]. However, reflection mode is not suitable for image sensor applications.

This chapter presents a nanoscale thick CMY color filter built from a hexagonal array of Al–TiO_2–Al nanorods on a quartz substrate that is derived from a subtractive MDM (metal-dielectric-metal) nanohole array structure. The structure exhibits a high transmission efficiency and narrow bandwidth to produce superior color separation.

Color tuning is achieved by varying the rod radius across the nanorod array, while keeping the base thickness constant, thereby making a mosaic of CMY filters across the substrate. An optical filter mosaic of size 4 mm × 4 mm was integrated onto a commercial monochrome image sensor to make a CMY camera, and its imaging capabilities demonstrated using the 12 color Macbeth chart.

4.2 Results

4.2.1 Filter Design

Figure 4.1a shows the proposed structure of the CMY color filter. The color filters are made of Al–TiO$_2$–Al nanorods fabricated on a quartz substrate and then embedded in a SOG (spin on glass [30]) matrix for refractive index matching.

The theory behind this CMY filter mosaic is based on Fano resonance, where the top and bottom metal disk work as dipole and cooperate to enhance the magnetic field

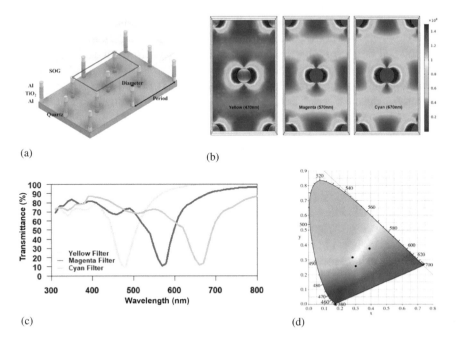

Fig. 4.1 a Simulation results of Al–TiO$_2$–Al nanorod based CMY (cyan, magenta, yellow) color filter mosaic: **a** Al–TiO$_2$-Al nanorods in hexagonal array on a quartz substrate covered with spin-on-glass, **b** the normalized electric field at valley wavelength for the filters, yellow (470 nm), magenta (570 nm) and cyan (670 nm) Electric field at 470nm, **c** numerically simulated transmission spectra of the CMY color filters. The wavelength is swept from 300 to 800 nm, **d** CIE chart of simulated CMY colors in the filter mosaic

Table 4.1 Parameters for Al–TiO$_2$–Al nanorod

	Cyan	Magenta	Yellow
Top Al thickness (nm)	40	40	40
TiO$_2$ thickness (nm)	90	90	90
Bottom Al thickness (nm)	40	40	40
Period (nm)	500	430	350
Diameter (nm)	60	90	40

between them and produce the Fano resonance. The spectral sharpness (the Q factor) relates to the ratio of the resonant frequency and the FWHM. It has been found that even though the upper wavelength resonant mode is sharper than the mode at smaller wavelength, most power is transferred by the second mode which requires symmetry within the MDM structure [27, 31]. Thus it is desirable to keep the thickness and radius of top metal disk the same as the bottom one. The peak wavelength in this structure depends primarily on the thickness of the metal disk and radius, but not the thickness of dielectric, which makes fabrication possible in a single stage without requiring multiple masks for each color.

The CMY filters were computationally investigated using finite element methods (FEM) implemented in COMSOL MULTIPHYSICS. The simulation model consists of a unit cell on a semi-infinite thick quartz substrate consisting of a single nanorod at the centre and one-quarter of nanorod at each corner, as shown in Fig. 4.1a. The simulation unit on the semi-infinite glass substrate is covered with spin-on-glass. Perfectly matched layers (PML) parameters were applied to the top and bottom layers and periodic boundary conditions on the four side of the simulation unit, as highlighted in the red block of Fig. 4.1a. For calculating the peak wavelengths, the refractive index of glass is taken as 1.5, spin of glass 1.45 and the wavelength dependent refractive index of Al from Rakic's data [32]. TiO$_2$ was deposited using an E-beam evaporator, the refractive index experimentally measured as 2.1 and this value used in the simulation model. The simulated transmission spectrum for CMY color filters are shown in Fig. 4.1c. The normalised electric field of each color filter, yellow (470 nm), magenta (570 nm) and cyan (670 nm) at their resonant wavelength (valley wavelength) is shown in Fig. 4.1b. The layer thickness, period and rod diameter are summarised in Table 4.1. Figure 4.1d shows that the CIE chromaticity chart of the simulated CMY color filters values fall within the appropriate part of the color space, thereby illustrating their suitability for imaging applications.

4.2.2   Color Mosaic Fabrication

A 4 mm × 4 mm CMY color mosaic was fabricated with an arrangement of CMYM color unit on a quartz substrate, repeating in both horizontal and vertical directions

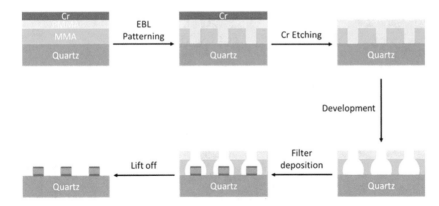

Fig. 4.2 Fabrication strategy for making Al–TiO$_2$–Al nanorod based CMY color filter mosaic

to create an array. The fabrication strategy using a PMMA-MMA bilayer is shown in Fig. 4.2. Firstly, a 1 mm thick quartz substrate was cleaned with acetone under sonication for five minutes followed by IPA and DI water rinse. Then the substrate was preheated on a heater under 80° for 10 min and then a thin EL9 (MMA) layer was spin coated on the quartz at 3000 rpm for 1.5 min and baked under 180° heater for 15 min. Next a thin PMMA A2 layer was spin coated onto the sample at 3000 rpm for 1.5 min and baked at 180° for 5 min. Due to the charging problem of quartz substrate with E-beam lithography (EBL), a 30 nm Cr layer was deposited on the sample by conformal sputtering to finish the sample preparation. After patterning with EBL (Vistec EBPG5000plusES), the Cr layer was removed using 1 min in Cr etchant, stopped by 5% H$_2$SO$_4$. The sample was then developed with diluted MIBK (MIBK:IPA = 1:3) for 1 min and stopped by IPA and DI water. The various profiles created in the PMMA and MMA layers after development are shown in Fig. 4.2. The bowl shape in the MMA, for example, will make the later lift off easier due to the smaller contact area between the Al–TiO$_2$–Al nanorod and quartz. Lastly, the Al–TiO$_2$–Al nanorod was deposited by an E-Beam evaporator (Intlvac Nanochrome II), note that the deposition rate of TiO$_2$ is 0.1 nm/s. After the lift off, the CMYM color filter was completed after being spin coated with SOG (Desert NDG-2000) at 2000 rpm for 20 s and baked at 210 °C for 15 min.

Figure 4.3d shows optical images of the CMYM filter mosaic in transmission mode under an optical microscope (Olympus BX53M) with 20× magnification. An enlarged image of the CMYM color mosaic is shown in Fig. 4.3e. The SEM image of the filter mosaic (cyan) from the top view is shown in Fig. 4.3f. Figure 4.3g shows in reflection mode that the reflected light from the sample (Fig. 4.3e) is RGB color. The transmission spectrum from the fabricated CMY color mosaic was measured using a CRAIC spectrometer (Apollo Raman^{TM} Microspectrometer) as shown in Fig. 4.3h. From the spectrum measurement, it can be seen the maximum transmission of Cyan, Magenta and Yellow color filter could be 90%. The transmission spectrum displayed high transmission without any secondary resonance from UV to Near-IR

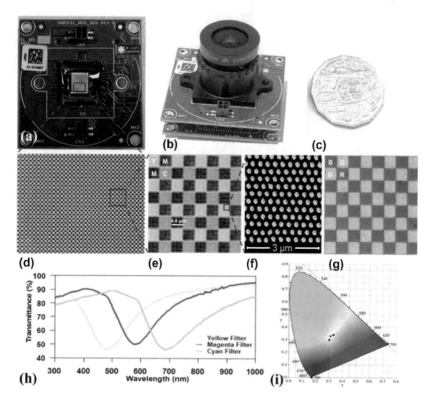

Fig. 4.3 Integration of the CMY color mosaic on the image sensor. **a** CMOS image sensor integrated with 4 mm × 4 mm CMY color mosaic using flip chip bonder for alignment, **b** CMY camera developed with optics (f number 1.4) for imaging, **c** a coin at the same scale as a reference to illustrate the compact size of the CMY camera, **d** the CMY color mosaic under optical microscope, **d** magnification ×20, **e** magnification ×40, **f** SEM image of the filter mosaic made of Al–TiO$_2$– Al nanorods from top view (cyan is shown), **g** reflected RGB colors from (**e**), **h** experimental transmission spectra of the CMY color filters from the color mosaic, **i** CIE chromaticity chart of the CMY filters in the mosaic from experimental data

wavelengths. Figure 4.3i shows the experimentally measured CIE chromaticity chart of the CMY color filters, and the detailed results are collated in Table 4.2.

From Fig. 4.3h, the valley wavelength in the simulation is almost the same as that in the experiments. It can be seen that the maximum transmission contrast is reduced, which also occurred in other experiments—e.g., [27]. This difference may be caused by the surface roughness of the deposited TiO$_2$. To keep a more consistent metal oxide property that results in a relatively high refractive index, the deposition rate of the E-beam evaporator needs to be as high as possible, causing a rough surface (around 3–5 nm measured by Atomic Force Microscopy). Therefore, the top Al disk becomes similarly rough. However, the maximum transmission efficiency in the experimental measurement is nearly the same as the simulation result, which

Table 4.2 Optical characteristics of the CMY filter mosaic

	Cyan		Magenta		Yellow	
	Simulation	Experiment	Simulation	Experiment	Simulation	Experiment
Valley wavelength (nm)	660	670	580	580	470	480
Maximum transmission contrast (%)	60	45	70	47	80	50
Maximum transmission efficiency (%)	90	90	90	90	90	90

can give us a large optical intensity, and the influence of the smaller transmission contrast on the color imaging applications is considered minimal. Furthermore, the contrast can be corrected in the later image processing steps.

4.2.3 Filter Integration and Image Reconstruction

The CMYM mosaic was integrated on a CMOS sensor using a flip-chip bonder (Fineplacer @Lambda) and captured raw images for further analysis. The CMOS image sensor used in this experiment is a monochrome sensor with a pixel size of 2.2 µm. Figure 4.3a shows that CMY color filter mosaic is aligned and glued on the image sensor, the glue is homemade with PMMA powder diluted with anisole solution.[2] The high transmission efficiency of the CMY color filter is demonstrated using the University logo on white paper, showing it can be seen clearly through the filter (Fig. 4.4). The whole filter mosaic size is 4 mm × 4mm and each color filter covers 2 by 2 pixels to increase the light absorption and decrease the spatial crosstalk. Therefore, the filter color pixel size is 4.4µm. The image sensor integrated with the mosaic is then fitted with a lens system with an f number of 1.4 for imaging as shown in Fig. 4.3b. The compact size of the CMY camera is shown with respect to a coin (shown at the same scale) in Fig. 4.3c.

This CMY camera is then used for the color imaging applications shown in Fig. 4.5. Firstly, a raw image of the standard 12-color Macbeth chart was taken with CMY camera (Fig. 4.5a). The recorded 12–bit raw image data was then transferred to a laptop for image processing using MATLAB. The raw data was analysed for saturation and signal intensity variation. Each color pixel was processed from the raw image to form three different images for cyan, magenta and yellow (Fig. 4.5b). Figure 4.5c shows the reconstructed CMY color image. Next, a CMY to RGB conversion algorithm was applied to change the CMY image to standard RGB image Fig. 4.5d: $R = Y + M - C$, $G = Y + C - M$, $B = C + M - Y$ [12]. Lastly, color

---

[2] The PMMA powder was diluted in a small amount of anisole, then left in a 100 class clean room for two weeks to amalgamate and reach the consistency of an adhesive.

Fig. 4.4 The fabricated CMY filter mosaic on the quartz substrate was cut into 4 mm × 4 mm using a dicing saw. The University of Melbourne logo on the white paper can be seen clearly through the filter. This demonstrate the high transmission through the filter mosaic, suitable for image sensor applications

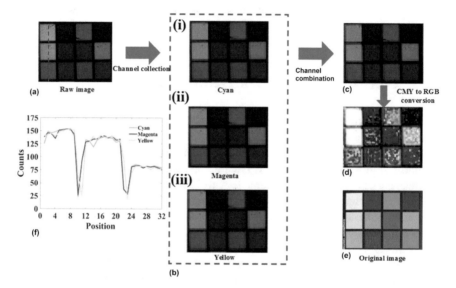

Fig. 4.5 Demonstration of color imaging by the CMY camera: **a** 12-bit raw image of the 12-color Macbeth chart captured by CMY camera. **b** Three bands (cyan, magenta and yellow) extracted from the raw image. **c** The three bands (three channel combination) are recombined to get a CMY color image. **d** The CMY color image is converted into RGB and applied color correction and white balance. **e** Standard image of the 12-color Macbeth color chart. **f** The plot shows signals from three pixel numbers along the vertical red dotted line in the raw image

correction and white balance were applied to recover the Macbeth chart as shown in Fig. 4.5e. Figure 4.5f shows a plot of the pixel intensity in the CMY raw image from the 12-color Macbeth chart across the transect indicated by the red dashed line. The results obtained demonstrates that colors can be retrieved from the 12-bit raw image data and that CMY camera is operating as expected.

4.3 Summary

This chapter has demonstrated a CMY camera built around a polarization independent nanoscale CMY color filter mosaic made of Al–TiO$_2$–Al nanorods exhibiting high transmission. A 4 mm × 4 mm filter mosaic has been integrated on a MT9P031 CMOS image sensor using flip-chip bonder and in-house prepared glue. Each filter color element is 4.4 μm arranged in a CMYM configuration. The performance of the filter mosaic was characterised and the CMY camera then evaluated using a 12-color Macbeth chart as an object and colors were retrieved using image processing algorithms. Although it is clear that the CMY filter array is capable of generating a color image, the use of more powerful image processing algorithms (as used in commercial cameras) will greatly improve its color performance. The presented technology will overcome the limitations imposed by conventional color filter technology for making next generation submicron pixels based CMY color image sensors and cameras, and will have applications in areas such as astronomy and low exposure time imaging in biology as well as general photography.

References

1. X. He, Y. Liu, P. Beckett, H. Uddin, A. Nirmalathas, R.R. Unnithan, CMY camera using a nanorod filter mosaic integrated on a CMOS image sensor. OSA Continuum **4**(1), 229–238 (2021)
2. S. Yokogawa, S.P. Burgos, H.A. Atwater, Plasmonic color filters for CMOS image sensor applications. Nano Lett. **12**(8), 4349–4354 (2012)
3. S.P. Burgos, S. Yokogawa, H.A. Atwater, Color imaging via nearest neighbor hole coupling in plasmonic color filters integrated onto a complementary metal-oxide semiconductor image sensor. ACS Nano **7**(11), 10038–10047 (2013)
4. K. Diest, J.A. Dionne, M. Spain, H.A. Atwater, Tunable color filters based on metal-insulator-metal resonators. Nano Lett. **9**(7), 2579–2583 (2009)
5. Y. Yan, Q. Chen, L. Wen, H. Xin, H.-F. Zhang, Spatial optical crosstalk in CMOS image sensors integrated with plasmonic color filters. Opt. Express **23**(17), 21994–22003 (2015)
6. R. Rajasekharan, E. Balaur, A. Minovich, S. Collins, T.D. James, A. Djalalian-Assl, K. Ganesan, S. Tomljenovic-Hanic, S. Kandasamy, E. Skafidas, D.N. Neshev, P. Mulvaney, A. Roberts, S. Prawer, Filling schemes at submicron scale: development of submicron sized plasmonic colour filters. Sci. Rep. **4**(1), 6435 (2014)
7. X. He, N. O'Keefe, Y. Liu, D. Sun, H. Uddin, A. Nirmalathas, R.R. Unnithan, Transmission enhancement in coaxial hole array based plasmonic color filter for image sensor applications. IEEE Photon. J. **10**(4), 1–9 (2018)
8. H.S. Song, G.J. Lee, D.E. Yoo, Y.J. Kim, Y.J. Yoo, D.-W. Lee, V. Siva, I.-S. Kang, Y.M. Song, Reflective color filter with precise control of the color coordinate achieved by stacking silicon nanowire arrays onto ultrathin optical coatings. Sci. Rep. **9**(1), 3350 (2019)
9. Q. Chen, D. Chitnis, K. Walls, T.D. Drysdale, S. Collins, D.R.S. Cumming, CMOS Photodetectors integrated with plasmonic color filters. IEEE Photon. Technol. Lett. **24**(3), 197–199 (2012)
10. B.Y. Zheng, Y. Wang, P. Nordlander, N.J. Halas, Color-selective and CMOS-compatible photodetection based on aluminum plasmonics. Adv. Mater **26**, 6318–6323 (2014)

11. R. Rajasekharan Unnithan, M. Sun, X. He, E. Balaur, A. Minovich, D. N. Neshev, E. Skafidas, A. Roberts. Plasmonic colour filters based on coaxial holes in aluminium. Mater. (Basel) **10**(4) (2017)
12. Astrosurf. RGB versus CMY color imagery. http://www.astrosurf.com/buil/us/cmy/cmy.htm. Accessed Jan 2021
13. T.W. Ebbesen, H.J. Lezec, H.F. Ghaemi, T. Thio, P.A. Wolff, Extraordinary optical transmission through sub-wavelength hole arrays. Nature **391**(6668), 667–669 (1998)
14. Active Silicon Inc. Lucky Star Gazing–High-Speed Imaging in Astronomy. https://www.activesilicon.com/news-media/news/lucky-star-gazing-high-speed-imaging-in-astronomy. Accessed July 2021
15. X. He, N. O'Keefe, D. Sun, Y. Liu, H. Uddin, A. Nirmalathas, R. Rajasekharan Unnithan, Plasmonic narrow bandpass filters based on metal-dielectric-metal for multispectral imaging, in *CLEO Pacific Rim Conference 2018* (Optical Society of America, 2018), p. Th4E.5
16. H.A. Bethe, Theory of diffraction by small holes. Phys. Rev. **66**, 163–182 (1944)
17. C. Genet, T. Ebbesen, Light in tiny holes. Nature. **445**, 39–46 (2007)
18. J. Zhang, M. ElKabbash, R. Wei, S.C. Singh, B. Lam, C. Guo, Plasmonic metasurfaces with 42.3% transmission efficiency in the visible. Light Sci. Appl. **8**(53) (2019)
19. V.R. Shrestha, S.-S. Lee, E.-S. Kim, D.-Y. Choi, Aluminum plasmonics based highly transmissive polarization-independent subtractive color filters exploiting a nanopatch array. Nano Lett. **14**(11), 6672–6678 (2014)
20. M. Ye, L. Sun, X. Hu, B. Shi, B. Zeng, L. Wang, J. Zhao, S. Yang, R. Tai, H.-J. Fecht, J.-Z. Jiang, D.-X. Zhang, Angle-insensitive plasmonic color filters with randomly distributed silver nanodisks. Opt. Lett. **40**, 4979–4982 (2015)
21. C. Yang, W. Shen, J. Zhou, X. Fang, D. Zhao, X. Zhang, C. Ji, B. Fang, Y. Zhang, X. Liu, L.J. Guo, Angle robust reflection/transmission plasmonic filters using ultrathin metal patch array. Adv. Opt. Mater. **4**, 1981–1986 (2016)
22. Y. Wu, A. Hollowell, C. Zhang, L.J. Guo. Angle-insensitive structural colours based on metallic nanocavities and coloured pixels beyond the diffraction limit. Sci. Rep. **3**(1194) (2013)
23. W. Yue, S. Gao, S. Lee, E.-S. Kim, D.-Y. Choi, Subtractive color filters based on a silicon-aluminum hybrid-nanodisk metasurface enabling enhanced color purity. Sci. Rep. **6**(29756) (2016)
24. F. Cheng, J. Gao, L. Stan, D. Rosenmann, D. Czaplewski, X. Yang, Aluminum plasmonic metamaterials for structural color printing. Opt. Express **23**, 14552–14560 (2015)
25. C. Yang, W. Shen, Y. Zhang, H. Peng, X. Zhang, X. Liu, Design and simulation of omni-directional reflective color filters based on metal-dielectric-metal structure. Opt. Express **22**, 11384–11391 (2014)
26. D. Fleischman, L.A. Sweatlock, H. Murakami, H. Atwater, Hyper-selective plasmonic color filters. Opt. Express **25**, 27386–27395 (2017)
27. H. Wang, X. Wang, C. Yan, H. Zhao, J. Zhang, C. Santschi, O.J.F. Martin, Full color generation using silver tandem nanodisks. ACS Nano. **11**(5), 4419–4427 (2017)
28. M.W. Knight, N.S. King, L. Liu, H.O. Everitt, P. Nordlander, N.J. Halas, Aluminum for plasmonics. ACS Nano. **8**(1), 834–840 (2014)
29. B. Zeng, Y. Gao, F.J. Bartoli, Ultrathin nanostructured metals for highly transmissive plasmonic subtractive color filters. Sci. Rep. **3**(1), 2840 (2013)
30. Desert Spin-On Glass. http://desertsilicon.com/spin-on-glass/. Accessed Jan 2021
31. T. Pakizeh, A. Dmitriev, M.S. Abrishamian, N. Granpayeh, M. Käll, Structural asymmetry and induced optical magnetism in plasmonic nanosandwiches. J. Opt Soc. Am. B **25**(4), 659–667 (2008)
32. A.D. Rakić, Algorithm for the determination of intrinsic optical constants of metal films: application to aluminum. Appl. Opt **34**(22), 4755–4767 (1995)

Chapter 5
A Single Sensor Based Multispectral Imaging Camera

A multispectral image camera captures image data within specific wavelength ranges in narrow wavelength bands across the electromagnetic spectrum. Images from a multispectral camera can extract additional information that the human eye or a normal camera fails to capture and thus may have important applications in precision agriculture, forestry, medicine and object identification. Conventional multispectral cameras are made up of multiple image sensors each fitted with a narrow pass-band wavelength filter and optics, which makes them heavy, bulky, power hungry and very expensive. The multiple optics also create image co-registration problem. This chapter demonstrates a single sensor based three band multispectral camera using a narrow spectral band RGB color mosaic in a Bayer pattern integrated on a monochrome CMOS sensor.[1] The narrow band color mosaic is made of a hybrid combination of plasmonic color filters and heterostructured dielectric multilayer. The resulting camera technology serves to reduce cost, weight, size and power by almost n times (where n is the number of bands) compared to a conventional multispectral camera.

5.1 Introduction

In a conventional CMOS based image sensor, color imaging relies on the integration of filters on top of the photodetector array [2–6]. These filters typically cover the three primary colors (bands), red, green and blue (RGB), predominately in a Bayer pattern [7, 8]. As the human eye is more sensitive to green light than either red or blue, the widely used Bayer filter mosaic is formed from alternating rows of red-

---

[1] Material in this chapter is reproduced from [1], licensed under a Creative Commons Attribution (CC BY 2.0).

X. He et al., *Multispectral Image Sensors Using Metasurfaces*, Progress in Optical Science and Photonics 17, https://doi.org/10.1007/978-981-16-7515-7_5

green and green-blue filters with twice as many green as red or blue filters. Three different materials are used for producing the primary colors with wide spectral bands (spectral width of around 90–100 nm) for all wavelengths [7, 9]. Multispectral cameras extend this concept to capture images with multiple color bands and with narrow pass bands (i.e., narrow spectral widths) [10–12]. Images from a multispectral camera can extract significant amount of additional information that the human eye or a normal camera fails to capture and thus have important applications in precision agriculture, forestry, medicine, object identification and classification [10–15].

Conventional multispectral cameras are made up of multiple CMOS sensors each externally fitted with a narrow passband filter. For example, three bands would require three image sensors with associated electronics, three narrow bandpass filters and three optical sub-systems. Depending on the application, the spectral width measured at the FWHM (Full Width Half Maximum) of a multispectral imaging camera may vary between 10 and 90 nm [10–12]. Figure 5.1 and Table 5.1 show comparisons between a conventional multispectral camera with six bands and a single sensor based multispectral camera.

The need for multiple sensors for each band can result in a number of problems. Firstly, it means that multispectral cameras tend to be bulky and power hungry, which in turn limits their wider deployment in portable applications such as drone-based precision agriculture or for hand- held and portable uses such as wound monitoring,

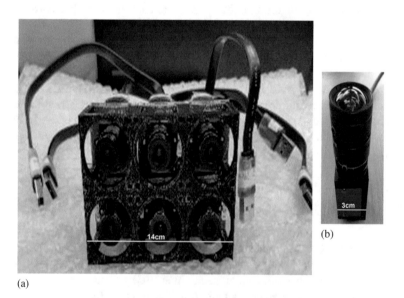

(a)

(b)

Fig. 5.1 Comparison between a conventional multispectral camera and single sensor based multispectral camera, **a** a compact conventional multispectral camera developed at electrical and electronic engineering, the University of Melbourne. The system has 6 different cameras, 6 bandpass filters, 6 processing electronics, 6 optics and bulky connectors, **b** single sensor based multispectral camera developed by the authors with the flexibility to integrate 3–6 bands

Table 5.1 Comparison between commercially available multispectral cameras and the proposed single sensor based multispectral camera

Parameter	Commercially available	Single sensor based
Size (L × W × D) (mm)	180 × 65 x 35	43 × 29 x 62
Weight (g)	520	130
Power consumption (W)	∼ 8	∼ 1.25
Image co-registration problems	Yes	No

vein detection and skin screening [13–15]. It also greatly complicates the task of optical alignment which aims to ensure that precisely the same scene is captured across all bands. Further, image co-registration problems will emerge from a slight mismatch between images in each band, which will require complex image processing to correct [16, 17]. These problems can be eliminated if a multispectral camera is developed using a single image sensor. This has prompted research in developing single sensor based multispectral cameras [18, 19].

Conventional pigments and dye-based filters are not suitable for making narrow band filters mosaic because their spectral widths tend to be too large (90–180 nm) and hence prevent the use of a normal color camera for multispectral imaging applications [11, 20]. Furthermore, these pigments tend also to be sensitive to UV radiation, degrade at high temperatures and are not particularly environmentally friendly.

Conventional technology for making a single narrow bandpass filter requires deposition of several layers of different dielectric materials (for example, 40 layers [21]) with precise thickness. In this case, making each narrow color band requires different thickness combinations for those 40 layers. For example, three narrow color bands require three different filters with each color band repeated a large number of times to form a filter mosaic on a CMOS sensor using conventional technology. This could conceivably require thousands of steps involving laying down thin films of precise thickness for each band separately with multiple complex masking and alignment processes, something that will likely result in a large failure rate, thereby significantly increasing manufacturing costs and complexity [22]. As a result, it is extremely difficult to develop a low cost color mosaic with narrow pass bands integrated onto the pixels of a CMOS chip.

Advances in nanofabrication has enabled the fabrication of novel nanophotonic devices including color filters [7, 22–26]. Plasmonic color filters with various geometries have proven their ability to tune the wavelength from UV to the NIR range [22, 27–48]. Furthermore, plasmonic filters have been shown to be suitable for developing narrow band filters in the visible wavelengths [46, 47] and short wavelength infrared (SWIR) [48]. In these filters, spectral FWHM values of 30 nm (visible) and 80 nm (SWIR) have been achieved in selected wavelengths. However, these filter technologies are not suitable for fabricating a narrow filter mosaic (i.e., comprising multiple narrow color bands next to each other) on a single substrate for multispectral imaging due to their complex architecture such as their requirement for different film thicknesses values for each band. 1-D metal grating filters has been explored with the aim to reduce the spectral width of plasmonic color filters. In 1-D metal grating filters, the plasmons are excited at the interface between the metal and dielectric [32, 46–48]. A spectral width of 64 nm was demonstrated using a 1-D metal grating filter made of silver nanoslit array fabricated on a glass substrate [48]. In addition, a 1-D metal grating filter based on Al nanoslit integrated onto a thick Al_2O_3 buffer layer has been demonstrated with spectral widths of 20 nm [32]. However, it has been proved both theoretically and experimentally [33, 49] that 1-D metal grating filters are polarization dependent and can produce colors only under transverse magnetic (TM) illumination. 2-D metal grating filters and dielectric guided mode resonance (GMR) filters are reported to avoid these polarization effects [26, 50, 51] and a hexagonal hole array inserted in a metal-dielectric-metal multilayer has been shown to slightly reduce the spectral width [22].

Overall, it appears that most of the narrow-band spectral filters reported are unsuitable for use in multispectral image sensors for a range of reasons, including an inability to achieve a narrow spectral width over a wide range of wavelengths, fabrication complexity to achieve narrow multi-band filters mosaic, polarization sensitivity and the requirement for TM illumination in the case of 1-D metal grating filters.

In this chapter, we demonstrate a single sensor based multispectral camera that employs a hybrid narrow spectral band RGB color mosaic fabricated on a quartz substrate and then integrated on a monochrome CMOS image sensor. The filter mosaic presented here can be easily tuned to any wavelength and requires only one processing stage to derive multiple bands. The filter mosaic consists of a hybrid combination of a double sandwich of silicon nitride-silica-silicon nitride layers (the heterostructured dielectric multilayer) covered by a hole array pattered in CMOS compatible aluminium (the plasmonic filter). The heterostructured dielectric multilayer is a common base layer for all the bands and the thickness values have been optimized to reduce the spectral width in a given wavelength range of interest. A single layer of plasmonic filter is used for wavelength tuning.

As a demonstration, the mosaic on quartz was integrated onto a commercial monochrome CMOS image chip using a flip-chip bonder resulting in a 3 cm × 3 cm size, three-band single-sensor based multispectral image camera. The performance of the camera was first demonstrated using a standard Macbeth Chart and was then fitted onto a lightweight DJI Phantom 3 drone to illustrate its capabilities for imaging in the field. Because only a single camera chip is required, weight, power and complexity can be reduced by a factor of n, where n is the number of bands compared to the conventional approach using multiple cameras, making this ideally suited to both airborne and handheld multispectral camera systems.

5.2 Hybrid Color Mosaic

In this section we describe the design, fabrication and performance of the RGB hybrid filter mosaic.

5.2.1 3-D Design and Optimization

The RGB hybrid filter mosaic was designed and optimized by 3-D simulation within the finite element based COMSOL Multiphysics® package. Each potential filter geometry was investigated separately using a 10 nm wavelength step size. Figure 5.2a shows the 3-D simulation model of the color mosaic with 6 layers. The basic unit cell for this simulation encompasses the diamond shaped pattern of holes highlighted in red in the top of Fig. 5.2a. The simulation model, Fig. 5.2a consisted of a 150 nm thick layer of aluminium patterned with a hexagonal arrangement of holes (the plasmonic filter: 1 layer) over repeating layers of Si_3N_4–SiO_2–Si_3N_4 (heterostructured dielectric multilayer: 5 layers), deposited on a semi-infinite glass substrate (n = 1.5). A 200 nm layer of Spin-On-Glass (SOG) was assumed to cover the aluminium layer for refractive index matching and to avoid any shorting with the metallic pads on the CMOS chip during physical integration. Finally, a perfectly matched layer (PML) was used at the top and bottom of the model to avoid the effects of reflected light in the transmittance spectrum and periodic boundary conditions were applied to the four sides to allow a large area to be simulated without costing excessive memory and time.

The pitch and hole diameters were varied to obtain peak transmission at 440, 530 and 625 nm [29] using the plasmonic layer. The objective was then to determine and validate the wavelength at which maximum transmittance occurs for these filters. Light was excited from the aluminium side (top side) using port boundary conditions and S-parameters used to find the transmittance ($|S21|^2$) of the filters. As in [50], refractive index values of 1.42 and 1.5 were used for the SOG layer and quartz substrate, respectively. A Filmetrics Model F20 was used to experimentally determine a refractive index for SiO_2 of 1.45 and around 1.9 for the Si_3N_4 and these values were then used in the simulations. As shown in Fig. 5.1b–d, the plasmonic layer has produced the required transmission peak (color), but with a large spectral width of 70 nm, 95 nm and 160 nm for blue, green and red respectively.

The plasmonic layer was then combined with a heterostructured dielectric multi-layer with 5 layers made of double Si_3N_4 (230 nm)–SiO_2 (350 nm)–Si_3N_4 (230 nm) sandwich layers to form a hybrid filter with optimized thickness values to reduce the spectral width. Here, photonic crystal theory is employed: if layer 1 has the refractive index n_1 and thickness l_1 and its adjacent layer 2 with n_2 and l_2, then when $l_1 n_1 = l_2 n_2$, the gap size will be maximized and therefore the incident light will produce a very narrow spectrum. Such an effect can be used to create narrow bandpass optical filters. The sharp peaks in the transmission spectrum as shown in Fig. 5.3 help to further narrow down the peak spectral width of the filters.

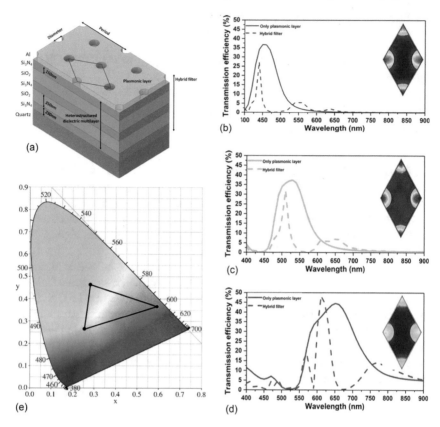

Fig. 5.2 a 3D simulation model of hybrid filter in the color mosaic. The hybrid filters consist of 6 layers made of double Si₃N₄ (230 nm)–SiO₂ (350 nm)–Si₃N₄ (230 nm) sandwich layers that forms a common base multilayer structure (heterostructured dielectric multilayer: 5 layers) and a 150 nm thick aluminium perforated with hexagonal arrangement of holes (plasmonic layer: 1 layer). **b** Numerically simulated transmission spectra of the blue plasmonic layer and blue hybrid filter showing the reduction in spectral width. The wavelength was swept from 400 to 900 nm. The spectral width at full width at half maximum (FWHM) of the hybrid blue filter was reduced to 17 from 70 nm produced by the plasmonic aluminum layer. The inset shows the normalized electric field at the peak wavelength of 440 nm, **c** the spectral width (FWHM) of hybrid green filter was reduced to 30 from 95 nm produced by the plasmonic layer. The inset shows the normalized electric field at the peak wavelength of 530 nm, **d** the spectral width (FWHM) of the hybrid red filter was reduced to 35 from 160 nm produced by the plasmonic layer. The inset shows the normalized electric field at the peak wavelength of 625 nm, **e** CIE chromaticity chart of simulated blue, green and red hybrid color filters in the mosaic

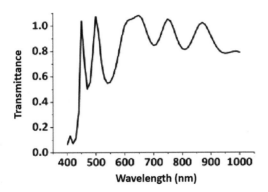

Fig. 5.3 Transmission spectrum of heterostructured dielectric multilayer

The heterostructured dielectric multilayer with a constant thickness forms a common base layer for all bands in the filter mosaic. In the hybrid filter, one single plasmonic layer removes most of the spectral contents on either side of the peak transmission wavelength in a given range of wavelengths. This eliminates the requirement for a large number of layers in the multilayer to produce narrow bands. Figure 5.2b–d shows the simulated spectra of the red, green and blue filters in the mosaic from 400 to 900 nm along with the electric field distributions at peak wavelengths. Here, the thickness of the Si_3N_4–SiO_2–Si_3N_4 base layer and the plasmonic layer (Al) are both kept constant and the pitch (period) and diameter of the holes in the 150 nm aluminium plasmonic layer are varied to tune the red, green and blue filters. The optimized hybrid filter mosaic parameters are given in Table 5.2.

Table 5.2 Optimized thickness parameters of the hybrid filters in the mosaic from computational simulation with COMSOL multiphysics® 5.2, the filter characteristics in the mosaic from simulations and experiments

Parameters (nm)		Red (620 nm)	Green (520 nm)	Blue (440 nm)
Common base structure: thickness kept constant	Si_3N_4 (layer 1)	230	230	230
	SiO_2 (layer 2)	350	350	350
	Si_3N_4 (layer 3)	230	230	230
	SiO_2 (layer 4)	350	350	350
	Si_3N_4 (layer 5)	230	230	230
	Aluminum (layer 6)	150	150	150
Pitch		420	340	290
Hole diameter		240	180	150
Simulated FWHM		35	30	17
Measured FWHM		45	60	60

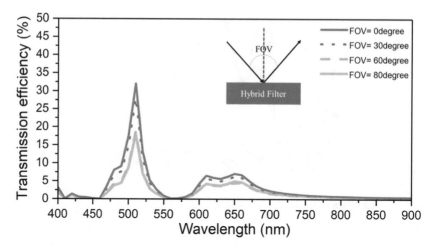

Fig. 5.4 Angle dependence of the hybrid filters. The green hybrid filter is taken as an example. The angle was varied from 0° to 80 ° FOV. The resonance peak position remains almost constant irrespective of the angle of incidence with slight decrease in the transmission intensity which is in the acceptable limit with suitable optics attached to the multispectral camera

The FWHM of the hybrid red filter was reduced to 35 nm in this simulated spectrum. For the green and blue filters, the FWHM was reduced to 30 nm and 17 nm respectively. Furthermore, this topology has considerably reduced the fabrication complexity as the thickness of the layers can be kept constant when fabricating the narrow band mosaic, and wavelength tuning can achieved by varying the pitch of the holes in the top single nanoscale thick plasmonic layer. The resonance peak shift with respect to different angle of incidence (0–80° field of view, FOV) was estimated for the hybrid filter. Figure 5.4 shows the green hybrid filter as an example. The resonant peak position remains almost constant irrespective of the angle of incidence with a slight decrease in the transmission intensity. This FOV will be within acceptable limits when suitable lenses are attached to the multispectral camera.

5.2.2 Fabrication Methods

The simplified fabrication process (on a 4-in. quartz wafer) is shown in Fig. 5.5. Firstly, the wafer was cleaned using Acetone and IPA (Isopropyl alcohol) with ultrasonic agitation followed by 2 min of plasma pre-cleaning. The wafer was then deposited with a-Si_3N_4 and a-SiO_2 by Plasma enhanced CVD (Oxford Instruments PLASMALAB 100 PECVD). The circular 4-in. wafer was diced into 2 cm × 2 cm square pieces and the centre pieces were selected for further fabrication due to the uniformity of the Si_3N_4 and SiO_2 film thicknesses. The measured refractive index of Si_3N_4 and SiO_2 developed by PECVD were 1.9 and 1.45, respectively. The deposition

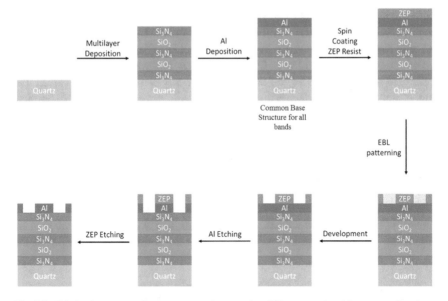

Fig. 5.5 Fabrication process for our proposed narrow band filters mosaic with common base

Table 5.3 Recipe for Si_3N_4 and SiO_2 deposition with PECVD

	RF forward power (W)	Chamber pressure (mT)	Temperature (° C)	5% SiH_4 and 95% Argon (Sccm)	NH_3 (Sccm)	N_2O (Sccm)	Deposition rate (nm/min)
Si_3N_4	20	1000	250	40	200	0	23
SiO_2	20	1000	250	150	0	800	70

rates were optimized and approximately 23 nm/min (10% tolerance) and 70 nm/min with the composition shown in Table 5.3.

The thickness of Si_3N_4 was optimized to be 230 nm and the SiO_2 to 350 nm. Starting with Si_3N_4, a total of five layers of Si_3N_4 and SiO_2 were deposited on the quartz wafer. After fabricating the multilayer structure, a 150 nm thick aluminium layer was deposited on the top of the structure using an E-beam evaporator (Intlvac Nanochrome II) at a constant rate of 0.2Å/s. An ellipsometer was used to measure the refractive index of the aluminium and the result showed it fits Rakić's experiment [52]. These data were subsequently used in the simulation model. A metallic nanohole array comprising varying pitch and hole diameters was fabricated on the aluminium film using EBL lithography process and deep reactive ion etching using the optimized values from simulations (Table 5.2). A thin ZEP 520A resist was spin coated on the device at 1500 rpm for 1.5 min, followed by 5 min baking at 180 °C. The pattern was exposed by EBL (Vistec EBPG5000plusES) with 1.5 nA current and 400 µm aperture for 4 h. The sample was then developed in n-Amyl acetate for a minute followed by a rinse with IPA and DI water. The exposed pattern was etched by deep reactive-ion etching (DRIE Oxford Instruments PLASMALAB100 ICP380) at 40 °C with forward power of 1000 W and 20 sccm Cl_2 under 2mT chamber pressure for 40 s to form the holes. The ZEP resist was then removed by DRIE at 40 °C with a forward power of 1000 W and 50 sccm O_2. Finally, the Spin-On-Glass [52] was spin coated on the top surface at 4000 rpm for 20 s, followed by baking on a hotplate at 210 °C for 10 min.

5.2.3   Spectrum Measurement and Discussion

The fabricated hybrid color filter on the quartz substrate was cut into 2 mm × 2 mm squares using a dicing saw with a G1A flange blade, and was optimized with hairline alignment. This alignment can adjust the cut on the substrate to the center of the

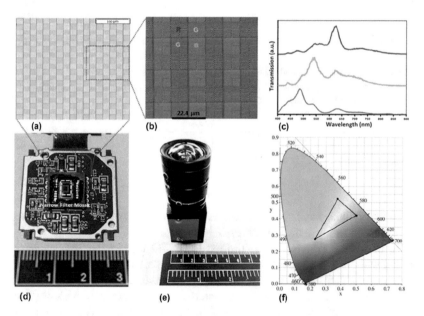

Fig. 5.6 Integration of the hybrid narrow spectral band mosaic on the image sensor **a** image of the color mosaic under optical microscope in transmission mode with magnification ×40, **b** SEM image of a section of the hybrid mosaic from top view. One narrow band RGBG unit size is 11.2 μm × 11.2 μm, **c** experimental transmission spectra of the narrow spectral band red, blue and green (RGB) color filters from the color mosaic. The spectral widths (FWHM) of RGB are 45 nm, 60 nm and 60 nm, respectively, **d** the color mosaic integrated on SONY ICX618 sensor pixels using a flip chip bonder for alignment (size 3 cm × 3 cm), **e** the single sensor based multispectral camera. The mosaic integrated image sensor fitted with housing and optics with f number 1.4 for multispectral imaging, **f** CIE chromaticity chart of the hybrid narrow band blue, green and red color filters in the color mosaic from experimental spectra

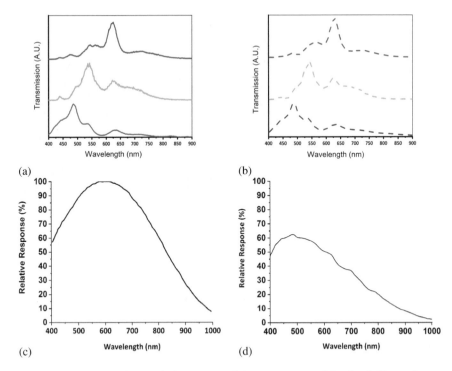

Fig. 5.7 **a** Experimental transmission spectra of the narrow spectral band red, blue and green (RGB) color filters from the color mosaic. The spectral widths of RGB are 45 nm, 60 nm and 60 nm, respectively, **b** the relative transmission spectrum after multiplying with image sensor response with respect to wavelength, **c** responsivity versus wavelength of ICX618, **d** responsivity versus wavelength of the MT9P031

hairline to precisely cut the edge with minimal edge damage, thus not affecting the sensor's optical performance. Figure 5.6a shows images of the RGB filter mosaic in transmission mode under an optical microscope (Olympus BX53M) with 40× magnification. SEM image of a section of the hybrid filter mosaic from top view is shown in Fig. 5.6b. One unit cell of the hybrid RGBG mosaic is 11.2 μm × 11.2 μm. The spectra of the hybrid RGB filters were measured using a CRAIC spectrometer (Apollo Raman™ Microspectrometer) and CytoViva Hyperspectral Imaging in transmission mode. Figure 5.6c shows the RGB spectrum from 400 to 900 nm. The spectral sensitivity (responsivity) of most commercial image sensors varies widely over their range in the visible to near-IR (Fig. 5.7). Hence the experimentally measured RGB spectra was multiplied by the transmission response versus wavelength for the image sensor to derive the actual spectra, which are shown in Fig. 5.7b. Figure 5.6f shows the CIE chromaticity chart overlaid with the transmission data, demonstrating that the RGB filter values are falling within the appropriate part of the color space. There is a small shift of green towards yellow due to a minor secondary peak in the green transmission spectrum. However, this small shift in the green coordinate still falls

Fig. 5.8 SEM images
showing fabrication variation

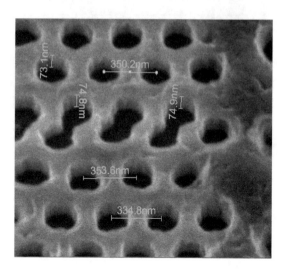

around the achromatic point and is within acceptable limits. The experimental transmission efficiency for RGB filters in the mosaic is around 10%. The low transmission
efficiency is compensated by making each filter band of size 11.2 μm covering 2 by
2 pixels (one pixel is 5.6 μm^2) to increase the light absorption by the photodetectors
and hence to increases the signal amplitude. Further compensation can be achieved
by increasing exposure time of the image sensor in low light conditions. The FWHM
of the red filter in the mosaic has the best performance with a width of around 45 nm,
while the blue and green filters both exhibit FWHM values of about 60 nm.

The measured FWHM values are slightly wider than the results obtained from
computer simulations for two primary reasons. Firstly, variations in the deposition
rate of Si_3N_4 coupled with the fact that SiO_2 growth using PECVD has larger tolerances than using the E-beam evaporator. While a high temperature (250 °C) during
the deposition of Si_3N_4 and SiO_2 results in a good quality dielectric, it restricts the
available methods for verifying the exact deposition thickness to the Filmetrics software sensor system in the PECVD, which is less accurate than AFM. Secondly, as
shown in the SEM image in Fig. 5.8, the pitch and the hole shape in the plasmonic
layer can vary due to fabrication tolerances (such as minor under-cut in holes and
nanoscale thickness variations) from the ideal (simulated) case. It can be seen that the
target pitch size is 340 nm, while the actual size varies from 334 to 353 nm. Further,
the shape of the hole is not a perfect circle. Together, these fabrication tolerances
result in the difference between experiment and simulations.

Color crosstalk among pixels is reduced by mounting the filter mosaic upside
down to minimize the effect of substrate thickness. This reduces the likelihood of
off-normal incident light of one pixel entering neighboring pixels. Further crosstalk
reduction was achieved by making each filter band of size 11.2 μm covering 2 by 2
pixels (one pixel size: 5.6 μm) to increase the light absorption by the photodetectors,
thus increasing the signal amplitude.

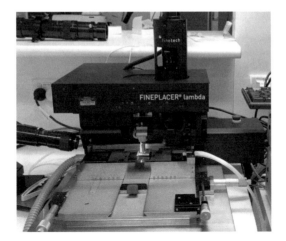

Fig. 5.9 Flip chip bonder (FINEPLACER@Lambda) used to align the filter mosaic on the image sensor

(a)

(b)

Fig. 5.10 **a** CMOS sensor with resolution 640 × 480 and a pixel size of 5.6 μm after removing the protective glass, **b** the color mosaic integrated on SONY CMOS sensor pixels using a flip chip bonder for alignment (size 3 cm × 3 cm)

The narrow band filter mosaic was then integrated onto a CMOS chip using a flip chip bonder (Fig. 5.9) for accurate alignment as shown in Fig. 5.6d. The top of the filter mosaic was coated with SOG to match the refractive index (thereby increasing the transmission) and also to reduce the spectral width. This also prevented the sensor being shorted during physical integration. The hybrid filter was integrated on the image sensor upside down to avoid crosstalk. The image sensor used was a SONY ICX618 with pixel size of 5.6 μm and resolution of around 0.3 Megapixels (640 × 480). The image sensor protective glass was removed for filter integration (Fig. 5.10).

To compensate for the low transmission of the filters and also to increase the light absorption in the photodetectors (pixels), each filter in the mosaic covered a 2 × 2

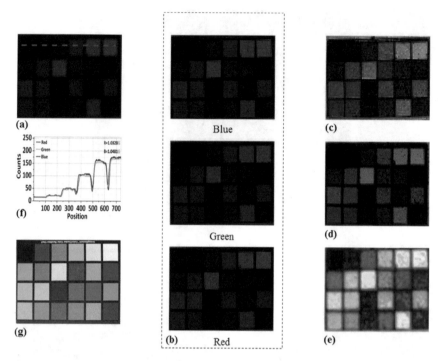

Fig. 5.11 Image reconstruction process from the single sensor based multispectral camera **a** 8 bit multispectral raw data of 24-patch Macbeth chart captured by the multispectral camera, **b** three narrow wavelength channels (RGB) extracted from the raw image, **c** the three channels are recombined to get a RGB color image, **d** the color image after color correction and **e** white balance, **f** the plot shows signals from pixel numbers along the dotted red line in the raw image, **g** original image of the 24-patch Macbeth color chart

block of photodetectors in the CMOS image sensor, resulting in 160 × 120 pixels per band. For integration, a PMMA based homemade adhesive was used[2] between the filter mosaic and the image sensor while performing the integration using the flip chip bonder. A thin layer of PMMA was spin coated on top of the image sensor and the filter integrated onto it after alignment with the flip-chip bonder (Fig. 5.9). The performance of the color filter was also verified by integrating on a second MT9P031 image sensor.

The optics used for the camera has f number 1.4 with a focal length of 6 mm. The single sensor based multispectral camera is shown in Fig, 5.6e. The camera was characterized using a 24-patch Macbeth Color Checker as an object (Fig. 5.11g). Eight bit multispectral raw object data was captured by the camera and then transferred to a laptop for image processing using a sequence of MATLAB steps as shown in Figs. 5.11a–e. As an example of the pixel intensity variations that are captured in the raw image, Fig. 5.11f shows a plot of the signals from the pixels across the transect

---

[2] The adhesive mixture preparation has been described previously in Sect. 4.2.3.

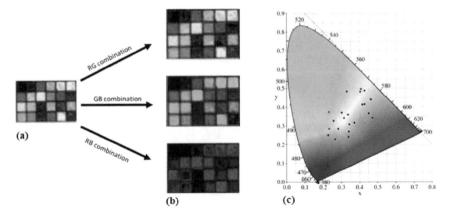

Fig. 5.12 Demonstration of image overlay of different bands **a** recovered Macbeth Chart from the single sensor based multispectral camera from Fig. 5.11, **b** RG, GB and RB combinations of Macbeth ColorChecker, **c** CIE chart for the recovered 24-patch Macbeth colors

indicated by the dashed red line on the Macbeth Chart in Fig. 5.11a. A demosaicing algorithm was used to extract red, blue and green channels from the multispectral raw data (Fig. 5.11b), then the red, green and blue channels were recombined to get a color image as shown in Fig. 5.11c. Due to the initial uncertainty of the RGB color balance, color correction and white balancing were required. The standard RGB spectrum could be treated as a specific color gamut such as sRGB, Adobe RGB and so on, each of which applies a particular interpretation for the values of the R, G and B channels for an 8-bit image. The three values of R, G and B for a given color in these gamuts (e.g., sRGB and Adobe RGB) will be different, resulting in a slightly different color output when applying the same image processing algorithms such as interpolation or demosaicing. It is apparent that the CIE chart for the fabricated RGB filter in this chapter is also different from these example spaces, and therefore requires color correction. Here, color balance and 'vibrance'[3] functions in the Photoshop® software were used to apply color correction and white balance. Figure 5.11d, e show the images after color correction and white balancing, respectively.

As illustrated by Fig. 5.11, each band can be retrieved from the 8-bit multispectral raw image data. Another requirement for a multispectral image camera applications is the extraction of each narrow band and their overlay into different spectral band combinations (red-green (RG), Green-Blue (GB) etc.).

Figure 5.12 shows that the recovered Macbeth Chart (Fig. 5.11e) can be used to recover different multispectral band combinations such as RG, RB and GB (Fig. 5.12b) which are useful in many applications. Examples include determining the NDVI (normalised differential vegetation index) for precision agriculture to find

---

[3] In this case, color balance was used to alter the overall mixture of colors used in the composite image. The Photoshop *Vibrance* function applies a non-linear adjustment to the saturation so that clipping is minimized as the colors approach full saturation. https://helpx.adobe.com.

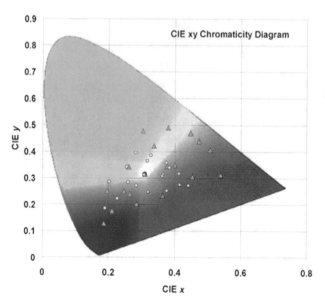

Fig. 5.13 CIE chart of a standard Macbeth chart. Triangles indicate the original Macbeth chart in the CIE chart. Circles indicate the recovered Macbeth chart in the CIE chart

plant diseases [20], finding required information in a band for object identification and also in finding emissions in a narrow band for biomedical applications. Figure 5.12c shows the CIE chart of the recovered Macbeth Chart. The chart demonstrates that the recovered color values are falling within the appropriate part of the color space in comparison to a standard CIE chart of the Macbeth Chart (Fig. 5.13).

The single sensor based multispectral camera was mounted on a DJI Phantom 3 for testing the sensor performance from a real aerial platform in an outdoor environment (an outer urban park) as shown in Fig. 5.14a, b. The sensor was mounted without a gimbal as shown in Fig. 5.14a. Figure 5.14c shows the raw multispectral images of the Macbeth color chart (printed on A0 paper) on the ground captured from a height of 15 m above the ground. The other patches are calibration images. Figure 5.14d shows the pixel intensity values across a line over the white and black crossing and the Macbeth Chart images captured by the multispectral camera. The pixel intensity variations with respect to positions are consistent with color intensity variations in the Macbeth Chart and black and color variations without any white balance. From the intensity variations, it could be seen that the image clarity was well within acceptable limits, and clearly indicates the difference across a range from black to grey and white color. Furthermore, the suitability of the raw multispectral image for making handheld sensors for precision agriculture was demonstrated by creating a Red-Blue vegetation index (RBVI), $(R - B) \times 255/(R + B)$. An area with both green and dry grass was captured by holding the drone mounted camera 1.5 m above the ground and the recording the raw image is shown in Fig. 5.14e. The individual R, G and B bands were recovered from the raw multispectral image to calculate the RBVI

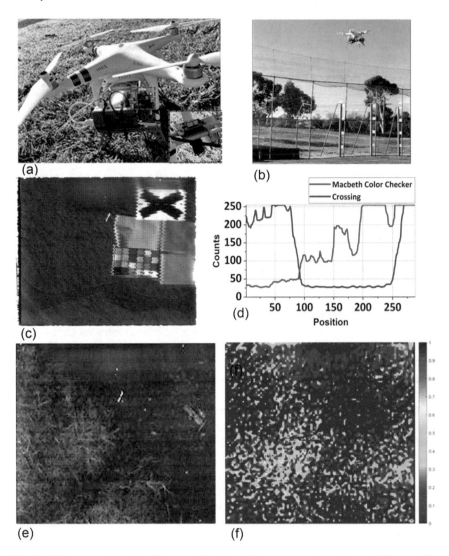

Fig. 5.14 Demonstration of multispectral imaging of the camera using a drone platform, **a** The single sensor based multispectral camera mounted on a DJI Phantom Drone without a gimbal, **b** image capturing using the camera, **c** 8 bit raw Image captured from 15 m elevation by the single sensor based camera showing the clarity of different patterns on ground, **d** the plot shows signals from pixel numbers along the dotted red line in the raw image of crossing (blue line) and Macbeth chart (red line), **e** 8 bit raw multispectral image of healthy grass and dry grass captured by the camera, **f** R, G and B bands are recovered from the raw multispectral image to get RB vegetation index (RBVI). RBVI images shows area of high dense green grass (red color in the image) compared to dry grass (blue color)

as shown in Fig. 5.14f. The RBVI image clearly shows areas of green grass (red color in the image) compared to dry grass (blue color) which demonstrates the capability of the sensor platform in a real application.

5.3   Summary

This chapter has demonstrated a single sensor based narrow band multispectral imaging using a hybrid RGB color mosaic integrated onto a CMOS sensor. The color mosaic was designed in such a way that multiple bands can be fabricated on a quartz wafer in a single run and offers easy tuning of colors, in contrast to conventional techniques that demand several independent runs with complex alignment processes. The hybrid filter mosaic was made of a heterostructured dielectric multilayer structure consisting of a Si_3N_4–SiO_2–Si_3N_4 sandwich as a common base layer for the filter mosaic to reduce the spectral width followed by a metal layer made of aluminum film perforated with holes on the base structure as a plasmonic layer. Color tuning is achieved by varying the pitch of holes and hence can be fabricated in a single run with no complex alignment required for the different bands. The thickness values required for the base and the plasmonic layers were optimized to obtain narrow spectral widths. The spectral widths of the RGB mosaic are 60 nm, 60 nm and 45 nm for the red, green and blue, respectively. The mosaic was then integrated onto a commercial sensor using a flip-chip bonder for better alignment accuracy with a thin layer of PMMA for adhesion and refractive index matching. The single sensor based narrow multispectral imaging capability was demonstrated using a Macbeth color chart followed by retrieving individual bands using demosaicing techniques and their combination to retrieve the Macbeth Chart after color correction and white balance. The sensor was then fitted onto a lightweight DJI Phantom 3 Drone to demonstrate its imaging capabilities in a field using RB Vegetation Index. Because only a single sensor chip was used for the camera, it required only around one third of the weight and power of a conventional multispectral camera. In general terms, the weight, power and complexity were reduced by a factor of n times, (where n is the number of bands) compared to a conventional multispectral camera using multiple sensors, electronics and optics. The sensor will be useful across a wide range of applications, including drone-based imaging for precision agriculture, developing portable low-cost sensors for wound healing, blood vein detection, mining, forensics and others.

As a final observation, the relatively low transmission measured in the device is due to the thick Al metal layer used in this particular design. Its value (150 nm) was selected to maintain a narrow FWHM after careful optimization via FEM simulation so is a necessary feature of the filter. However, there are several ways to increase the transmission efficiency while still maintaining a narrow FWHM and this aspect is discussed in Chap. 6.

References

1. X. He, Y. Liu, K. Ganesan, A. Ahnood, P. Beckett, F. Eftekhari, D. Smith, M.H. Uddin, E. Skafidas, A. Nirmalathas, R. Rajasekharan Unnithan, A single sensor based multispectral imaging camera using a narrow spectral band color mosaic integrated on the monochrome CMOS image sensor. APL Photon. **5**(4), 046104 (2020)
2. M. Miyata, M. Nakajima, T. Hashimoto, High-sensitivity color imaging using pixel-scale color splitters based on dielectric metasurfaces. ACS Photon. **6**(6), 1442–1450 (2019)
3. S. Goossens, G. Navickaite, C. Monasterio, S. Gupta, J.J. Piqueras, R. Pérez, G. Burwell, I. Nikitskiy, T. Lasanta, T. Galán, E. Puma, A. Centeno, A. Pesquera, A. Zurutuza, G. Konstantatos, F. Koppens, Broadband image sensor array based on graphene–CMOS integration. Nat. Photon. **11**(6), 366–371 (2017)
4. J. Tang, Q. Cao, G. Tulevski, K.A. Jenkins, L. Nela, D.B. Farmer, S. Han, Flexible CMOS integrated circuits based on carbon nanotubes with sub-10 ns stage delays. Nat. Electron. **1**, 191–196 (2018)
5. M. Bariya, H.Y.Y. Nyein, A. Javey, Wearable sweat sensors. Nat Electron. **1**, 160–171 (2018)
6. T. Sugie, T. Akamatsu, T. Nishitsuji, R. Hirayama, N. Masuda, H. Nakayama, Y. Ichihashi, A. Shiraki, M. Oikawa, N. Takada, Y. Endo, T. Kakue, T. Shimobaba, T. Ito, High-performance parallel computing for next-generation holographic imaging. Nat. Electron. **1**, 254–259 (2018)
7. R. Rajasekharan, E. Balaur, A. Minovich, S. Collins, T.D. James, A. Djalalian-Assl, K. Ganesan, S. Tomljenovic-Hanic, S. Kandasamy, E. Skafidas, D.N. Neshev, P. Mulvaney, A. Roberts, S. Prawer, Filling schemes at submicron scale: development of submicron sized plasmonic colour filters. Sci. Rep. **4**(1), 6435 (2014)
8. Z. Fan, H. Qiu, H. Zhang, X. Pang, L. Zhou, L. Liu, H. Ren, Q. Wang, and J.A Dong. Broadband achromatic metalens array for integral imaging in the visible. Light Sci. Appl. **8**(67) (2019)
9. S. Ortega, H. Fabelo, D.K. Iakovidis, A. Koulaouzidis, G.M. Callico, Use of hyperspectral/multispectral imaging in gastroenterology. Shedding some–different –light into the dark. J. Clin. Med. **8**(1), 36 (2019)
10. V. Lebourgeois, A. Bégué, S. Labbé, B. Mallavan, L. Prévot, B. Roux, Can commercial digital cameras be used as multispectral sensors? A crop monitoring test. Sensors **8**(11), 7300–7322 (2008)
11. H. Park, K.B. Crozier, Multispectral imaging with vertical silicon nanowires. Sci. Rep. **3**(1), 2460 (2013)
12. L. Duempelmann, B. Gallinet, L. Novotny, Multispectral imaging with tunable plasmonic filters. ACS Photon. **4**(2), 236–241 (2017)
13. C.-T. Pan, M. D. Francisco, C.-K. Yen, S.-Y. Wang, Y.-L. Shiue. Vein pattern locating technology for cannulation: a review of the low-cost vein finder prototypes utilizing near infrared (NIR) light to improve peripheral subcutaneous vein selection for phlebotomy. Sensors **19**(3573) (2019)
14. X. Ai, Z. Wang, H. Cheong, Y. Wang, R. Zhang, J. Lin, Y. Zheng, M. Gao, B. Xing, Multispectral optoacoustic imaging of dynamic redox correlation and pathophysiological progression utilizing upconversion nanoprobes. Nat. Commun. **10**(1087) (2019)
15. Y. Chang, H. Yicheng, Z. Chen, D. Xing, Co-impulse multispectral photoacoustic microscopy and optical coherence tomography system using a single supercontinuum laser. Opt. Lett **44**(18), 4459–4462 (2019)
16. X. Zhao, J. Zhang, C. Yang, H. Song, Y. Shi, X. Zhou, D. Zhang, G. Zhang, Registration for optical multimodal remote sensing images based on fast detection, window selection, and histogram specification. Remote Sens. **10**(663) (2018)
17. T. Skauli, An upper-bound metric for characterizing spectral and spatial coregistration errors in spectral imaging. Opt. Express **20**(2), 918–933 (2012)
18. IMEC. Hyperspectral imaging. https://www.imec-int.com/en/hyperspectral-imaging, 2021. Accessed June 2021
19. Spectral Devices Inc., Multispectral camera solutions. https://www.spectraldevices.com/, 2021. Accessed June 2021

20. X. He, N. O'Keefe, D. Sun, Y. Liu, H. Uddin, A. Nirmalathas, R. Rajasekharan Unnithan, Plasmonic narrow bandpass filters based on metal-dielectric-metal for multispectral imaging, in *CLEO Pacific Rim Conference 2018* (Optical Society of America, 2018), p. Th4E.5
21. M. Scalora, M.J. Bloemer, A.S. Pethel, J.P. Dowling, C.M. Bowden, A.S. Manka, Transparent, metallo-dielectric, one-dimensional, photonic band-gap structures. J. Appl. Phys. **83**(5), 2377–2383 (1998)
22. X. He, N. O'Keefe, Y. Liu, D. Sun, H. Uddin, A. Nirmalathas, R. Rajasekharan Unnithan, Transmission enhancement in coaxial hole array based plasmonic color filter for image sensor applications. IEEE Photon. J. **10**(4), 1–9 (2018)
23. R. Rajasekharan, H. Butt, Q. Dai, T.D. Wilkinson, G. Amaratunga, Can nanotubes make a lens array? Adv. Mater. **24**(23), OP170–OP173 (2012)
24. H. Butt, Q. Dai, R. Rajasekharan, T.D. Wilkinson, G. Amaratunga, Enhanced reflection from arrays of silicon based inverted nanocones. Appl. Phys. Lett. **99**(13), 133105 (2011)
25. R. Rajasekharan, Q. Dai, T.D. Wilkinson, Electro-optic characteristics of a transparent nanophotonic device based on carbon nanotubes and liquid crystals. Appl. Opt. **49**(11), 2099–2104 (2010)
26. Y. Ding, R. Magnusson, Resonant leaky-mode spectral-band engineering and device applications. Opt. Express **12**(23), 5661–5674 (2004)
27. M.J. Uddin, T. Khaleque, R. Magnusson, Guided-mode resonant polarization-controlled tunable color filters. Opt. Express **22**(10), 12307–12315 (2014)
28. R. Rajasekharan Unnithan, M. Sun, X.He, E. Balaur, A. Minovich, D.N. Neshev, E. Skafidas, A. Roberts, Plasmonic colour filters based on coaxial holes in aluminium. Mater. (Basel) **10**(4) (2017)
29. S.P. Burgos, S. Yokogawa, H.A. Atwater, Color imaging via nearest neighbor hole coupling in plasmonic color filters integrated onto a complementary metal-oxide semiconductor image sensor. ACS Nano **7**(11), 10038–10047 (2013)
30. X. Duan, S. Kamin, N. Liu, Dynamic plasmonic colour display. Nature. Communications **8**(1), 14606 (2017)
31. J. Sautter, I. Staude, M. Decker, E. Rusak, D.N. Neshev, I. Brener, Y.S. Kivshar, Active tuning of all-dielectric metasurfaces. ACS Nano **9**(4), 4308–4315 (2015)
32. D.B. Mazulquim, K.J. Lee, J.W. Yoon, L.V. Muniz, B.-H.V. Borges, L.G. Neto, R. Magnusson, Efficient band-pass color filters enabled by resonant modes and plasmons near the rayleigh anomaly. Opt. Express **22**, 30843–30851 (2014)
33. B. Zeng, Y. Gao, F.J. Bartoli, Ultrathin nanostructured metals for highly transmissive plasmonic subtractive color filters. Sci. Rep. **3**(1), 2840 (2013)
34. S. Yokogawa, S.P. Burgos, H.A. Atwater, Plasmonic color filters for CMOS image sensor applications. Nano Lett. **12**(8), 4349–4354 (2012)
35. Y.-T. Yoon, C.-H. Park, S.-S. Lee, Highly efficient color filter incorporating a thin metal-dielectric resonant structure. Appl. Phys. Express **5**(2), 022501 (2012)
36. D. Inoue, A. Miura, T. Nomura, H. Fujikawa, K. Sato, N. Ikeda, D. Tsuya, Y. Sugimoto, Y. Koide, Polarization independent visible color filter comprising an aluminum film with surface-plasmon enhanced transmission through a subwavelength array of holes. Appl. Phys. Lett. **98**(093113) (2011)
37. G. Si, Y. Zhao, H. Liu, S. Teo, M. Zhang, T.J. Huang, A.J. Danner, J. Teng, Annular aperture array based color filter. Appl. Phys. Lett **99**, 033105 (2011)
38. L.B. Sun, X.L. Hu, B.B. Zeng, L.S. Wang, S.M. Yang, R.Z. Tai, H.J. Fecht, D.X. Zhang, J.Z. Jiang, Effect of relative nanohole position on colour purity of ultrathin plasmonic substractive colour filters. Nanotechnology **26**(30) (2015)
39. H.-S. Lee, Y.-T. Yoon, S.-S. Lee, S.-H. Kim, K.-D. Lee, Color filter based on a subwavelength patterned metal grating. Opt. Express **15**(23), 15457–15463 (2007)
40. Q. Chen, D. Chitnis, K. Walls, T.D. Drysdale, S. Collins, D.R.S. Cumming, CMOS photodetectors integrated with plasmonic color filters. IEEE Photon. Technol. Lett. **24**(3), 197–199 (2012)

41. B.Y. Zheng, Y. Wang, P. Nordlander, N.J. Halas, Color-selective and CMOS-compatible photodetection based on aluminum plasmonics. Adv. Mater. **26**(36), 6318–6323 (2014)
42. T.W. Ebbesen, H.J. Lezec, H.F. Ghaemi, T. Thio, P.A. Wolff, Extraordinary optical transmission through sub-wavelength hole arrays. Nature **391**(6668), 667–669 (1998)
43. H.A. Bethe, Theory of diffraction by small holes. Phys. Rev. **66**, 163–182 (1944)
44. C. Genet, T. Ebbesen, Light in tiny holes. Nature. **445**, 39–46 (2007)
45. M. Ye, L. Sun, X. Hu, B. Shi, B. Zeng, L. Wang, J. Zhao, S. Yang, R. Tai, H.-J. Fecht, J.-Z. Jiang, D.-X. Zhang, Angle-insensitive plasmonic color filters with randomly distributed silver nanodisks. Opt **40**, 4979–4982 (2015)
46. D. Fleischman, L.A. Sweatlock, H. Murakami, H. Atwater, Hyper-selective plasmonic color filters. Opt. Express **25**, 27386–27395 (2017)
47. D. Fleischman, K.T. Fountaine, C.R. Bukowsky, G. Tagliabue, L.A. Sweatlock, H.A. Atwater, High spectral resolution plasmonic color filters with subwavelength dimensions. ACS Photon. **6**(2), 332–338 (2019)
48. Y. Liang, S. Zhang, X. Cao, L. Yanqing, X. Ting, Free-standing plasmonic metal-dielectric-metal bandpass filter with high transmission efficiency. Sci. Rep. **7**(1), 4357 (2017)
49. S.A. Maier, *Plasmonics—Fundamentals and Applications*, 1st edn. (Springer, 2007)
50. Desert Spin-On Glass. http://desertsilicon.com/spin-on-glass/. Accessed Jan 2021
51. W.-K. Kuo, C.-J. Hsu, Two-dimensional grating guided-mode resonance tunable filter. Opt. Express **25**(24), 29642–29649 (2017)
52. A.D. Rakić, Algorithm for the determination of intrinsic optical constants of metal films: application to aluminum. Appl. Opt. **34**(22), 4755–4767 (1995)

Chapter 6
Hybrid Color Filters for Multispectral Imaging

Multispectral cameras capture images in multiple wavelengths in narrow spectral bands. They offer advanced sensing well beyond normal cameras and many single sensor based multispectral cameras have been commercialized aimed at a broad range of applications, such as agroforestry research, medical analysis and so on. However, existing single sensor based multispectral cameras require accurate alignment to overlay each filter on the image sensor pixels, which makes their fabrication very complex, especially when the number of bands is large. This chapter uses computational simulations to analyze a new filter technology using a hybrid combination of a single plasmonic layer and dielectric layers. A filter mosaic of various bands with narrow spectral width can be achieved with single run manufacturing processes (i.e., exposure, development, deposition and other minor steps), regardless of the number of bands.[1]

6.1 Introduction

Multispectral camera (MC) technology has the ability to collect information well beyond human eyes or normal color cameras. This is because the human eye and conventional color cameras receive an image in three color bands—red, green and blue (RGB)—with broad spectral widths. Hence, it is difficult to distinguish fine spectral features within an image captured across these bands due to their overlap. For example, the spectral width in conventional cameras is around 80–100 nm for each of the RGB bands. Multispectral cameras based on a single image sensor capture images

---

[1] Text and diagrams in this chapter from [1] © Wiley-VCH GmbH. Reproduced with permission. This material may not be reproduced in any form except for accessible versions made by non-profit organizations serving the blind, visually impaired and other persons with print disabilities (VIPs).

© The Author(s), under exclusive license to Springer Nature Singapore Pte Ltd. 2021
X. He et al., *Multispectral Image Sensors Using Metasurfaces*, Progress in Optical Science and Photonics 17, https://doi.org/10.1007/978-981-16-7515-7_6

in multiple wavelengths using narrow band filters fitted on each image sensor pixel [2–5]. It is possible to extract additional spectral information otherwise not available from these multiple images [2–11]. The spectral width of each band is measured at FWHM of the spectral distribution, where FWHM is defined as the wavelength range at which the received optical power is half of its maximum.

Multispectral camera technology sub-divides the information collected by the image sensor into narrow bands defined by the small FWHM values of its constituent filters. The filtered information can then be grouped, processed and highlighted to show a large contrast with its reference background. Examples already abound across diverse applications. In gastroenterology research, each observed case (i.e., tissue type) may relate to a particular range of electromagnetic radiation or to a different reflectance spectrum. In this case, ulceration of the colon has been shown to exhibit a larger reflectance at the boundary between red and NIR wavelengths (around 600 nm to 1000 nm) [7] and can therefore be made to appear much brighter against a non-diseased background. In the pathology domain, the location of veins can be difficult to detect for many reasons, such as dark skin, excess loss of blood or even extreme dehydration. To solve this problem, MC been applied as either a combination of a commercial camera and a NIR bandpass filter or a standard camera together with NIR illumination. As NIR light will easily penetrate the skin, medical staff are able to more easily identify the exact position of veins [6]. In agroforestry research, it is known that diseased or stressed plants reflect more NIR light than healthy leaves. Thus, MC systems have been installed on drone aircraft to fly over the crops, observing harvest conditions in real time [8]. One example of this application was illustrated previously in Chap. 5.

There are three different types of MC system currently on the market. The first uses multiple cameras with a single narrow-band filter in front of each camera [9, 10]. This type of system can be easily customized simply by replacing the spectral filter with one appropriate to the specific application. The main disadvantage is that these multiple camera platforms tend to be bulky and power hungry, which limits their use [11]. Moreover, each camera has their own imaging viewpoint which may create co-registration problems [12, 13]. While an second type that employs a single camera plus a rotating optical filter disk can eliminate the viewpoint disparity problem, they are still quite bulky and introduce time shifts that complicate the processing of moving images, where the system is integrated with a drone for agroforestry applications, for example [10–14].

The third type of MC available on the market is based on a single camera sensor called single sensor based multispectral camera (SSMCs) [2–5]. Although the filter technologies employed in these systems are almost universally commercial-in-confidence, the limited information available in the open literature (e.g., [5]) indicates that the filter wavelength is typically tuned by changing the filter material and/or its thickness. This kind of technology is also described in many recent research papers [15, 16] and various patents [17–20].

The photodetector under each pixel collects the optical signal after it has passed through both a micro-lens and the filter layer(s) [21]. There are typically two ways to install a filter array on top of an image sensor. It can be either fabricated onto a substrate and then aligned with the image sensor, or the array can be directly fabricated onto the surface of the sensor itself. Either method requires a large number of separate fabrication processes for each individual spectral band, such as pixel-to-pixel mask alignment, UV exposure, and filter installation. Therefore, the fabrication of the filter mosaic with required number of bands can become extremely complicated, particularly in the case of hyperspectral filters where the number of filters (bands) may reach into the hundreds [5, 11].

This has triggered research on new filter technologies suitable for SSMCs. The two main requirements here are developing technologies with less complex fabrication processes and achieving narrow spectral width (FWHM) for each band. There has been much activity aimed at improving the FWHM of the bands, as evidenced by recent patents and published papers [15, 17–35]. By far the smallest FWHM reported to date in the visible/NIR range is approximately 0.11 nm observed over a very limited 10 nm wavelength range in a dielectric grating structure [29]. A metallic grating structure has also been demonstrated in which an aluminium grating built on an Al_2O_3 buffer layer, in turn located on a quartz substrate, results in a minimum FWHM of 20 nm [31]. Metal-Dielectric-Metal (MDM) or even MDMDM metallic gratings have also been designed and optimized to exhibit good FWHM values of under 30 nm [32]. However, as these metallic grating structures have been shown to be polarization dependent, SSMCs based on them may fail to register an optical signal if the incident light is in the opposite polarization direction.

Metallic hole array based multispectral filters have been proposed [11, 24, 28, 29, 33–37] that show FWHM value smaller than 50 nm. Although these multispectral filters have been applied to color and multispectral imaging applications [11], they are unable to operate over the whole visible and NIR range. Si nanowire based narrow bandpass filter arrays have also been integrated on a monochrome image sensor for multispectral imaging, but with mechanical scanning and have a FWHM greater than 50 nm [30].

In this chapter, we demonstrate a new filter technology suitable for multispectral imaging using a hybrid combination of a single plasmonic layer and dielectric multilayers. The hybrid filters can produce spectral width (FWHM) values smaller than 50 nm, with a minimum FWHM as low as 17 nm at some wavelengths. Moreover, these proposed multispectral filters are easily tuneable and a multiband filter mosaic can be manufactured using a single stage maskless fabrication process regardless of the number of spectral bands, including one-time exposure, chemical development and filter deposition. We have demonstrated two filter structures as examples, one exploiting localized surface plasmons resonances and other using surface plasmon resonances employing the proposed hybrid topology to demonstrate their operation in the visible and NIR wavelengths by computational simulations.

6.2 Design of Hybrid Multispectral Filters

6.2.1 Angle Independent Narrow Bandpass Filters Based on Localized Surface Plasmon Resonances

In the first of the proposed filters presented here, Aluminium (Al) disk structures were investigated. In a previous chapter, these were used to develop subtractive color filtering as the different resonant wavelengths (valley wavelengths) depend largely on the disk diameter [38]. Subtractive primary colors typically include cyan, magenta and yellow, and are the inverse transmission spectra of red, green and blue respectively. The work described in this chapter has resulted in a multispectral filter mosaic (RGB color filter plus NIR filter) using Al disks in a square array based on a hybrid technique: Al disc (plasmonic layer using localized surface plasmons) plus dielectric layers (silicon nitride (Si_3N_4) and spin-on-glass (SOG)).

An initial narrow-band filter was designed and optimized using finite element method (FEM) implemented in the simulation software COMSOL Multiphysics 5.5. The device geometries were designed and simulated in 3-D using the radio-frequency module (RF module) in COMSOL Multiphysics. Periodic boundary condition (PBC) and semi-infinite boundary condition (perfect electric conductor and perfect magnetic conductor) were used for studying the effects of infinite and semi-infinite filter size. Perfectly Matched layers were used to suppress any unwanted reflections in the simulation model. Scattering boundary conditions were used to terminate the simulation boundaries where reflection effects are minimal. The transmission coefficient is obtained from S-parameters using the port boundary conditions. It was found that by appropriately arranging the Al disks in a square array as shown in Fig. 6.1a, the peak transmission wavelength could be tuned by varying the diameter of the Al disk, while the optimal spacing between two adjacent disks was found to be around 1/10th

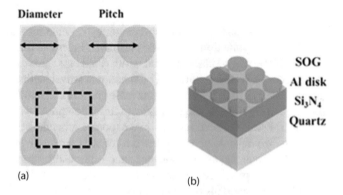

Fig. 6.1 Schematic of the proposed narrow-band filter based on the hybrid Al disk **a** top view of the filter (dotted line shows the unit cell used in the simulation), **b** 3-D representation of the filter structure with the plasmonic layer (Al disc) and dielectric layers (Si_3N_4 and SOG)

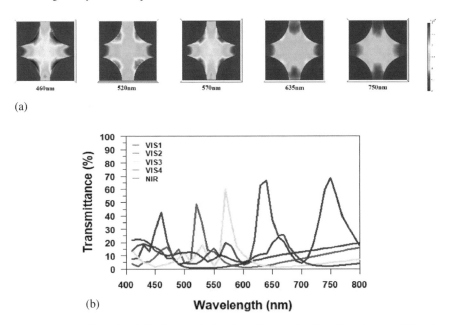

Fig. 6.2 **a** Normalized electric field of localised surface plasmons for blue (460 nm), green (520 nm), yellow (570 nm), red (635 nm) and NIR (750 nm), **b** transmission spectra of hybrid Al disk based multispectral filters with respect to the wavelengths in (**a**)

of the peak wavelength. In these simulations, the substrate is quartz with a refractive index of 1.5. A 250 nm thick Si_3N_4 layer with a refractive index of 2 was deposited as a buffer layer on the quartz. A thin plasmonic layer of 50 nm Al disks with refractive index given in [39] was added to the buffer layer. This was designed to tune the peak wavelength. Lastly, a 200 nm coating of Spin on glass (SOG) covered the sample. The general 3-D structure of the filter is shown in Fig. 6.1b. PMLs were applied at the top and bottom, SBCs layers to the remaining sides of the simulation model. Port boundary conditions with propagation constants of $2\pi \times 1.42/\lambda$ for SOG and $2\pi \times 1.5/\lambda$ for quartz were used between the top PML and SOG as well as between the bottom PML and the quartz. Periodic boundary conditions (PBC) were applied on the four sides of the middle Al disk (unit cell as shown within the dotted area in Fig. 6.1a) and Si_3N_4 block.

The transmission efficiency was calculated using the S-parameter, $|S21|^2$. Figure 6.2a shows the normalized electric field at peak wavelengths for blue (460 nm), green (520 nm), yellow (570 nm), red (635 nm) and NIR (750 nm) showing the localized surface plasmons. The resulting FWHM for green and yellow (Fig. 6.2b) are about 18 nm and 15 nm, respectively, with transmission efficiency values larger than 45%. The wavelength tuning was mainly achieved by tuning diameter of the disk (Fig. 6.2a). Table 6.1 shows the optimized filter parameters along with their optical characteristics.

Table 6.1 Results of blue, green, yellow, red and NIR narrow bandpass filters

Parameter	Blue	Green	Yellow	Red	NIR
Peak wavelength (nm)	460	520	570	635	750
Period (nm)	240	280	310	350	420
Disk diameter (nm)	200	220	260	280	340
Peak transmission (%)	42	48	60	65	68
FWHM (nm)	25	18	15	30	45

Figure 6.3a shows the CIE chromaticity chart overlaid with the transmission data of blue, green and red showing that the filter values are falling in the appropriate part of the color space. The transmission efficiency with respect to angle of incidence was investigated using the green filter (520 nm) as shown in Fig. 6.3b. The full field of view (FFOV) values varied from 0° (i.e., normal to the surface) to 80° off-axis. It was observed that the peak wavelength position does not shift with respect to the angle and exhibits a minimal reduction in the transmission intensity up to 30% off normal. When the FFOV is 80°, the green filter may adversely affect the optical response of the NIR. As has been seen previously from Fig. 4.1 in Chap. 4, the proposed hybrid filter geometry is made of Al nano-disks plus a Si_3N_4 layer, and exhibits FWHM values smaller than 50 nm in the visible and NIR wavelength. It is therefore a promising candidate for making multispectral filter mosaics. When the peak wavelength is smaller than 600 nm, the FWHM reduces to about 15 nm, which is suitable for making multispectral cameras with a large number of bands. Moreover, the proposed narrow-band filter behavior is independent of the angle of incidence and, as all the structures (substrate, Si_3N_4, Al, SOG) have the same base thickness, the array requires only a single-exposure fabrication process without any mask alignment.

6.2.2 Metallic Nanohole Array Integrated on a Dielectric Multilayer for IR Multispectral Imaging

The second of the two structures presented in this chapter is a hybrid filter illustrated in Fig. 6.4 using surface plasmon resonances. This filter is made up of a plasmonic filter on a dielectric multilayer on top of a quartz substrate (refractive index 1.5) covered with Spin on Glass with refractive index 1.42 (Fig. 6.4). A dielectric multilayer is formed from a layer of SiO_2 with refractive index (n_1) of 1.45 surrounded by two TiO_2 layers with larger refractive indexes (n_2) of 2.1. The thickness of the

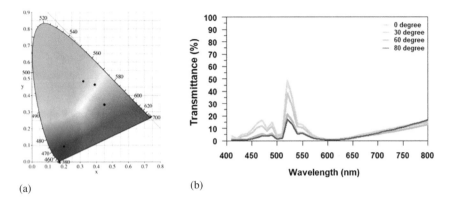

(a)   (b)

Fig. 6.3 **a** CIE chromaticity chart for green, blue, yellow and red, **b** transmission versus wavelength over FFOV: $0° \leq$ angle of incidence $\leq 80°$

Fig. 6.4 Nanostructure of the proposed hybrid IR multispectral filter **a** top view, **b** 3-D structure of the filter structure (Al-Plasmonic layer and TiO_2, SiO_2, SOG are dielectric layers)

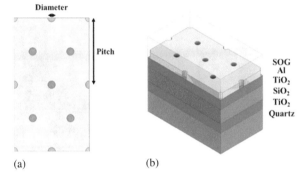

(a)   (b)

TiO_2 layers are both 190 nm, which is 25% of the target NIR wavelength of 760 nm. The SiO_2 layer is governed by the function $d_1 n_1 = d_2 n_2$, where d_1 and d_2 are the thicknesses of the SiO_2 and TiO_2 layers, respectively, and n_1 and n_2 are the corresponding refractive indices, resulting in a SiO_2 thickness of approximately 280 nm. This part of the filter on its own behaves as an anti-reflection layer with a passband from approximately 650–1000 nm as illustrated by the transmission spectrum (Fig. 6.5) of the dielectric multilayer before the plasmonic layer is added. The plasmonic filter comprises an array of nano-holes in a hexagonal arrangement [40] inserted in the 150 nm Au film. Figure 6.5a shows the normalized electric field of the surface plasmons in the hole array (top view) at peak wavelengths for NIR 1-5.

The transmission efficiency and the resulting spectrum of the hybrid filter is plotted in Fig. 6.5b. The inset of Fig. 6.5b shows the spectrum from TiO_2-SiO_2-TiO_2 dielectric multilayers without the plasmonic layer. The FWHM and peak transmission efficiency results for the device are presented in Table 6.2. As shown in this table, the FWHM of a multispectral filter in the NIR wavelength can be less than 25 nm (typically around 20 nm) with a best-case transmission efficiency of 80%. The dielectric multilayer remains the same in each case, showing that the peak wavelength can be

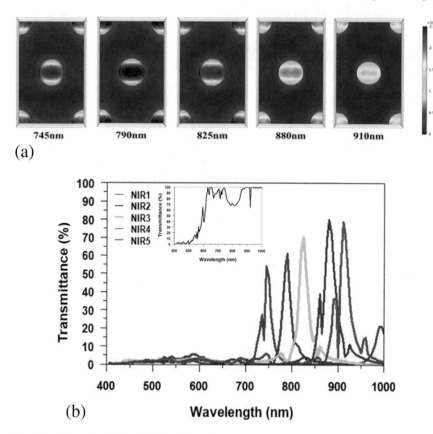

Fig. 6.5 **a** Normalized electric field of the surface plasmons in the hole array at peak wavelengths for NIR 1–5, **b** The simulated transmission spectrum of the hybrid IR multispectral filters. The inset shows the transmission spectrum of TiO$_2$–SiO$_2$–TiO$_2$ dielectric multi-layer without the plasmonic layer

Table 6.2 Results of Au hole array integrated with TiO$_2$–SiO$_2$–TiO$_2$ multilayer

Peak wavelength (nm)	745	790	825	880	910
Period (nm)	480	520	550	600	630
Hole diameter (nm)	180	200	220	240	240
Peak transmission efficiency (%)	55	60	70	80	78
FWHM (nm)	20	20	22	25	20

tuned by simply adjusting the period between adjacent holes. However, hole array based filters are limited in their angle independence due to the SPPs. Color filters based on LSPs are more angle independent than those that are SPP-based.

We have carried out further simulations to study the effect of non-infinite filter sizes on the transmission coefficient and spectral width by replacing the periodic boundary conditions with semi-infinite boundary conditions. There is a clear difference in minimum operating dimensions evident between the LSP and SPP mechanisms. In the case of the metallic disk based structures, which operate via LSP, structures as small as even a single disk will produce the same color as a larger array of disks. It is therefore possible to reduce the size of a filter based on metal disks to below hundreds of nanometers, while still achieving similar transmission efficiencies and spectral widths.

In contrast, the transmission efficiency of a SPP based color filter depends on its size, so it is not possible to greatly reduce this below a certain level. As one example, filters based on the nano-hole arrays described above rely on interference between surface plasmons and hence multiple holes are required to produce a color. Thus, the spectral width and transmission efficiency for these types of hole array filters reduce as the filter size decreases. Further, because the filter blocks are intended to align with the underlying pixel array in the image sensor, this implies a minimum area for the pixels as well. Given that the commercial sensors used in the experiments here and in Chap. 5 have pixel sizes of around 2.2–2.4 µm it was found that to produce the required spectral width along with acceptable transmission efficiency, the hole array based filter had to cover at least 2×2 pixels on the image sensor, resulting in a final pixel size of about 4.8µm. In contrast, LSP operation can support filter sizes well below the pixel dimensions in current state of the art image sensors.

6.3 Summary

Using computational simulations, this chapter has analysed a multispectral filter technology using a hybrid combination of single plasmonic layer and dielectric layers . The hybrid filter technology reduces the fabrication complexity to a single stage process regardless of the number of bands in a filter mosaic. The filters also exhibit narrow spectral widths and easy wavelength tuneability.

We have demonstrated the operation of the proposed hybrid design in the visible and NIR using two filter examples, one using localized surface plasmons resonances and other surface plasmon resonances. This proposed methodologies will significantly reduce the complexity and cost of fabricating multispectral filter mosaics compared to existing complex SSMC filter techniques. This is likely to lead to SSMCs being more widely applied in areas such as precision agriculture, medicine, forestry, night vision and remote sensing. In the future, there will almost certainly be a demand for a greater number of bands and should be increased and spectral width is expected to be reduced to less than 10 nm for making low cost single sensor based hyper spectral cameras.

References

1. X. He, Y. Liu, P. Beckett, M.H. Uddin, A. Nirmalathas, R. Rajasekharan Unnithan, Hybrid color filters for multispectral imaging. Adv. Theory Simul. **3**(11), 2000137 (2020)
2. IMEC. Hyperspectral imaging. https://www.imec-int.com/en/hyperspectral-imaging, 2021. Accessed June 2021
3. Spectral Devices Inc. Multispectral camera solutions. https://www.spectraldevices.com/, 2021. Accessed June 2021
4. Ocean Insight. Products. https://www.oceaninsight.com/products/, 2021. Accessed June 2021
5. Ximea Inc. Hyper spectral imaging. https://www.ximea.com/. Accessed June 2021
6. C.-T. Pan, M.D. Francisco, C.-K. Yen, S.-Y. Wang, Y.-L. Shiue. Vein pattern locating technology for cannulation: a review of the low-cost vein finder prototypes utilizing near infrared (NIR) light to improve peripheral subcutaneous vein selection for phlebotomy. Sensors **19**(3573) (2019)
7. S. Ortega, H. Fabelo, D.K. Iakovidis, A. Koulaouzidis, G.M. Callico, Use of hyperspectral/multispectral imaging in gastroenterology. Shedding some-different-light into the dark. J. Clin. Med. **8**(1), 36 (2019)
8. U. Mahajan, Drones for normalized difference vegetation index (NDVI). https://support.dronedeploy.com/docs/ndvi-cameras-for-drones. Accessed June 2021
9. Tetracam Inc. Imaging systems. https://www.tetracam.com/ImagingSystems.htm, 2020. Accessed June 2021
10. G. Themelis, J.S. Yoo, V. Ntziachristos, Multispectral imaging using multiple-bandpass filters. Opt. Lett **33**, 1023–1025 (2008)
11. X. He, Y. Liu, K. Ganesan, A. Ahnood, P. Beckett, F. Eftekhari, D. Smith, M.H. Uddin, E. Skafidas, A. Nirmalathas, R. Rajasekharan Unnithan, A single sensor based multispectral imaging camera using a narrow spectral band color mosaic integrated on the monochrome CMOS image sensor. APL Photon. **5**(4), 046104 (2020)
12. T. Skauli, An upper-bound metric for characterizing spectral and spatial coregistration errors in spectral imaging. Opt. Express **20**(2), 918–933 (2012)
13. L.L. Coulter, D.A. Stow, Assessment of the spatial co-registration of multitemporal imagery from large format digital cameras in the context of detailed change detection. Sensors **8**(4), 2161–2173 (2008)
14. L. Yao, G. Xiaozhou, Y. Zhong, Z. Han, Q. Shi, F. Ye, C. Liu, X. Wang, T. Xie, Image enhancement based on in vivo hyperspectral gastroscopic images: a case study. J. Biomed. Opt. **21**(10), 101412 (2016)
15. S. Pimenta, S. Cardoso, A. Miranda, P. De Beule, E.M.S. Castanheira, G. Minas, Design and fabrication of SiO_2/TiO_2 and Msenmaier, agO/TiO_2 based high selective optical filters for diffuse reflectance and fluorescence signals extraction. Biomed. Opt. Express **6**(8), 3084–3098 (2015)
16. Y.-J. Jen, A. Lakhtakia, M.-J. Lin, W.-H. Wang, W. Huang-Ming, H.-S. Liao, Metal/dielectric/metal sandwich film for broadband reflection reduction. Sci. Rep. **3**(1), 1672 (2013)
17. D.H Cushing. Multilayer thin film dielectric bandpass filter. U.S. Patent US5926317A, 20 Aug 1999 [Online]. Available https://patents.google.com/patent/US6018421A/pt
18. R.-Y. Tsai, H.-Y. Lin, Y.-H. Chen, C.-S. Chang, Polarization-independent ultra-narrow band pass filters. U.S. Patent US20020080493A1, 27 June 2003 [Online]. Available https://patents.google.com/patent/US6018421A/pt
19. K.L. Lewis, Multilayer optical filters. U.S. Patent US6631033B1, 7 Oct 2003 [Online]. Available https://patents.google.com/patent/US6631033B1/en?oq=US6631033B1
20. A.C. Kundu, Multilayer band pass filter. U.S. Patent US7312676B2, 25 Dec 2007 [Online]. Available https://patents.google.com/patent/US7312676B2/en?oq=U.S +Patent+No.+7312676B2
21. Y. Huo, C.C. Fesenmaier, P.B. Catrysse, Microlens performance limits in sub-2 μm pixel CMOS image sensors. Opt. Express **18**(6), 5861–5872 (2010)

22. Y.-J. Jen, C.-C. Lee, L. Kun-Han, C.-Y. Jheng, Y.-J. Chen, Fabry-Pérot based metal-dielectric multilayered filters and metamaterials. Opt. Express **23**(26), 33008–33017 (2015)

23. R. Rajasekharan, E. Balaur, A. Minovich, S. Collins, T.D. James, A. Djalalian-Assl, K. Ganesan, S. Tomljenovic-Hanic, S. Kandasamy, E. Skafidas, D.N. Neshev, P. Mulvaney, A. Roberts, S. Prawer, Filling schemes at submicron scale: Development of submicron sized plasmonic colour filters. Sci. Rep. **4**(1), 6435 (2014)

24. X. He, N. O'Keefe, Y. Liu, D. Sun, H. Uddin, A. Nirmalathas, R. Rajasekharan Unnithan, Transmission enhancement in coaxial hole array based plasmonic color filter for image sensor applications. IEEE Photon. J. **10**(4), 1–9 (2018)

25. R. Rajasekharan Unnithan, M. Sun, X. He, E. Balaur, A. Minovich, D. N. Neshev, E. Skafidas, A. Roberts, Plasmonic colour filters based on coaxial holes in aluminium. Mater. (Basel), **10**(4) (2017)

26. X. He, Y. Liu, P. Beckett, H. Uddin, A. Nirmalathas, R. Rajasekharan Unnithan, Transmission enhancement in plasmonic nanohole array for colour imaging applications, vol. 11200, in *AOS Australian Conference on Optical Fibre Technology (ACOFT) and Australian Conference on Optics Lasers, and Spectroscopy (ACOLS)*, eds. by A. Mitchell, H. Rubinsztein-Dunlop (International Society for Optics and Photonics, SPIE, 2019), pp. 227–228

27. X. He, N. O'Keefe, D. Sun, Y. Liu, H. Uddin, A. Nirmalathas, R. Rajasekharan Unnithan, Plasmonic narrow bandpass filters based on metal-dielectric-metal for multispectral imaging, in *CLEO Pacific Rim Conference 2018* (Optical Society of America, 2018), p. Th4E.5

28. Y. Liang, S. Zhang, X. Cao, L. Yanqing, X. Ting, Free-standing plasmonic metal-dielectric-metal bandpass filter with high transmission efficiency. Sci. Rep. **7**(1), 4357 (2017)

29. S. Tibuleac, R. Magnusson, Narrow-linewidth bandpass filters with diffractive thin-film layers. Opt. Lett. **26**(9), 584–586 (2001)

30. H. Park, K.B. Crozier, Multispectral imaging with vertical silicon nanowires. Sci. Rep. **3**(1), 2460 (2013)

31. D.B. Mazulquim, K.J. Lee, J.W. Yoon, L.V. Muniz, B.-H.V. Borges, L.G. Neto, R. Magnusson, Efficient band-pass color filters enabled by resonant modes and plasmons near the rayleigh anomaly. Opt. Express **22**, 30843–30851 (2014)

32. D. Fleischman, K.T. Fountaine, C.R. Bukowsky, G. Tagliabue, L.A. Sweatlock, H.A. Atwater, High spectral resolution plasmonic color filters with subwavelength dimensions. ACS Photon. **6**(2), 332–338 (2019)

33. H.-S. Lee, Y.-T. Yoon, S.-S. Lee, S.-H. Kim, K.-D. Lee, Color filter based on a subwavelength patterned metal grating. Opt. Express **15**(23), 15457–15463 (2007)

34. Q. Chen, D. Chitnis, K. Walls, T.D. Drysdale, S. Collins, D.R.S. Cumming, CMOS photodetectors integrated with plasmonic color filters. IEEE Photon. Technol. Lett. **24**(3), 197–199 (2012)

35. L. Duempelmann, B. Gallinet, L. Novotny, Multispectral imaging with tunable plasmonic filters. ACS Photon. **4**(2), 236–241 (2017)

36. S. Yokogawa, S.P. Burgos, H.A. Atwater, Plasmonic color filters for CMOS image sensor applications. Nano Lett. **12**(8), 4349–4354 (2012)

37. S.P. Burgos, S. Yokogawa, H.A. Atwater, Color imaging via nearest neighbor hole coupling in plasmonic color filters integrated onto a complementary metal-oxide semiconductor image sensor. ACS Nano **7**(11), 10038–10047 (2013)

38. M. Ye, L. Sun, X. Hu, B. Shi, B. Zeng, L. Wang, J. Zhao, S. Yang, R. Tai, H.-J. Fecht, J.-Z. Jiang, D.-X. Zhang, Angle-insensitive plasmonic color filters with randomly distributed silver nanodisks. Opt. Lett. **40**, 4979–4982 (2015)

39. A.D. Rakić, Algorithm for the determination of intrinsic optical constants of metal films: application to aluminum. Appl. Opt. **34**(22), 4755–4767 (1995)

40. M. Sun, M. Taha, S. Walia, M. Bhaskaran, S. Sriram, W. Shieh, R. Rajasekharan Unnithan, A photonic switch based on a hybrid combination of metallic nanoholes and phase-change vanadium dioxide. Sci. Rep. **8**(1), 11106 (2018)

Chapter 7
Conclusions and Future Outlook

This book has presented the development of a new low-cost single sensor based multispectral image sensor camera technology exploiting metasurface techniques. This has addressed some of the major limitations of conventional multispectral cameras, by eliminating the requirement to set up multiple image sensors, their associated electronics, and collimating optics for each band, which can otherwise cause high power consumption and image co-registration problems due to slight mismatches between the images in each band.

It is already clear that metasurfaces will greatly impact the field of optics and imaging, especially in the case of multispectral imaging, where optical filters must exhibit high transmission and extremely sharp passbands across the optical and near infra-red spectra. To this end, Chap. 1 introduces the basic concepts behind metasurfaces for optical applications as these relate to the creation of meta lenses, polarized image sensors and colour filter applications. Chapter 2 then extends these concepts to discuss the characteristics of metasurfaces for multispectral imaging with a special focus on colour filters reported in the literature.

Chapter 3 commences with a novel strategy developed using computational simulations to overcome the challenge with low transmittance in plasmonic coaxial hole array based color filters without compromising their advantages, such as polarization and incident angle independence. In the future, with more advanced nanotechnology available, this kind of nanostructure will be a good candidate to be further researched, developed and fabricated, as it combines the advantages of both localized surface plasmons and surface plasmon polaritons for fine tuning of resonances, suppressing unwanted peaks and for increasing the transmission percentage.

A new nanorod-based cyan, magenta and yellow (CMY) colour filter technology is presented in Chap. 4 aimed at the next generation of image sensors with submicron pixel sizes. The CMY is a subtractive colour scheme with better performance in low light conditions compared to the corresponding RGB system. Conventional CMY colour filters use the absorption characteristics of their component pigments and

X. He et al., *Multispectral Image Sensors Using Metasurfaces*, Progress in Optical Science and Photonics 17, https://doi.org/10.1007/978-981-16-7515-7_7

dyes and hence they cannot be fabricated at nanoscale dimensions. To address this limitation, a color filter mosaic made of CMOS compatible Al–TiO$_2$–Al nanorods has been designed and fabricated and then integrated onto a commercial CMOS image sensor to demonstrate its feasibility in imaging applications. In the future new metasurfaces with enhanced absorption in a narrow wavelength range can be explored for use in the CMY colour space.

A color mosaic formed from of a hybrid combination of plasmonic color filters and heterostructured dielectric multilayers is discussed in Chap. 5. This combined filter structure achieves a narrow spectral width well suited to building a single sensor based multispectral image camera. The imaging capability of the camera has been characterised and tested, initially using a standard Macbeth chart then in an outdoor environment by fitting it to a lightweight DJI Phantom 3 drone. This technology will greatly reduce the cost, weight, size and power of multispectral cameras compared to current approaches that require multiple optics and image sensors. The basic concept could be usefully extended in the future by increasing its transmission using emerging low loss metasurfaces and by further reducing the spectral widths and greatly increasing the number of bands to enable 'hyper-spectral' imaging.

Chapter 6 uses computational techniques to further investigate the behavior of colour filters formed using a hybrid combination of a single plasmonic surface and dielectric layers. The focus of this study was to explore two filter topologies, one using localized surface polaritons (LSP) and other based just on surface plasmon resonances. The objective was to increase the spectral imaging range from the visible to near infrared wavelengths and at the same time offer narrower spectral widths and reduced fabrication complexity. These experiments clearly show that metallic nanostructure-based filters, which operate via LSP, can behave as narrowband filters even if reduced to structures as small as even a single object (e.g., a nano-disk). These single nano-structures will produce the same color filtering effects along with similar transmission efficiencies and spectral widths as a larger array. At the time of writing, pixel sizes have already reached 0.7 µm[1] and market trends appear to be driving this towards sub-wavelength pixel dimensions below 0.6 µm. Clearly, the ability to create simple, highly efficient narrow filters at sub-micron dimensions will become increasingly important into the future.

Multispectral imaging has already proved to be vital for remote sensing applications from space based platforms. However, the weight, size and power of the conventional spectral imaging systems are constraints in using them for the next generation miniaturised satellites, such as those known as "CubeSats". The single sensor based technology promises to considerably reduce the weight, size and power requirements with increased imaging capability and so will be well suited for deployment in these small satellites. Such developments may be complemented with new image processing algorithms that can compensate for the very long imaging distances (hundreds of kilometers) that can reduce the resolution of the single sensor

---

[1] From 2020, Samsung 5G smartphones incorporate a front-facing ("selfie") 48MP image sensor with 700 nm pixel pitch. Source: https://semiengineering.com/scaling-cmos-image-sensors.

based spectral imaging systems. The resolution issues can be reduced if high resolution images sensors are used for such applications with distortion-corrected optical systems.

Multispectral imaging has enormous applications in biomedical imaging and early diagnosis of diseases. The images in different wavelengths can distinguish features and biological changes in living organisms otherwise not possible to distinguish from the bulk tissue. Compared to traditional techniques, spectral imaging has been found to be more accurate in evaluating the depth of skin burns and their subsequent healing. Furthermore, in-vivo spectral imaging has already been demonstrated to be a powerful tool for the diagnosis of diseases such as oral cancer. However, such exploration is limited now due to the fact that the existing multispectral systems are bulky and not readily portable. The single sensor based multispectral imaging will support the development of handheld portable devices that will expand the use of multispectral imaging in the medical field.

Although the multispectral camera can record an enormous amount more information compared to a conventional camera, the full information retrieval is still challenging for some practical applications where a definite detection/classification is not possible. In the future, it is likely that this will be able to be addressed using machine learning (ML) approaches and artificial intelligence (AI). A promising direction is the use of trained models to design the colour filters based on metasurfaces for specific applications. Furthermore, the requirement to simultaneously achieve narrow spectral widths and high pixel resolution may be able to be relaxed if new image enhancement algorithms can be developed using machine learning and artificial intelligence techniques. For example, "super-resolution" (SR) is an algorithmic process already in common use to increase the resolution of diagnostic images. Applying these techniques to the multispectral domain can achieve higher quality results than conventional image re-sampling methods and may have the added advantage of at least partially removing the pressure driving further pixel miniaturization, thereby potentially lowering cost and complexity.

Another promising direction is the extension of the multispectral concepts from the visible region to mid-infrared (MWIR) and long wavelength infrared (LWIR) wavelengths. This can overcome the limited information capturing capability of the conventional 'thermal' image sensors and requires research on new metasurfaces operating in the thermal wavelengths.

While much progress has already been made, our ability to manipulate materials at an atomic scale is likely to open up brand new functionalities that are as yet unimaginable. In this book we have tried to illustrate some of the more promising directions. However, it is clear that we are just at the beginning of this journey into the realm of atomic scale materials and surfaces and there is still much work to be done to fully realise the opportunities offered by metasurfaces.

Appendix A
Simulation and Nanofabrication Methods

The software used in the work described in this book was COMSOL Multiphysics®, which employs finite element methods (FEM) [1] to simulate physical structures. For example, a 1-D function can be be divided into small pieces and represented with differential equations that interact with their neighboring functions to iterate to a stable solution for the overall geometry. The same method can be applied to 2-D and 3-D structures, resulting in a mesh of interacting solutions. COMSOL is also able to apply the same techniques to solve a wide range of problems across fields such as mechanical engineering, chemistry, even nuclear reactions. In this work, the radio frequency (RF) module was used to design the various filters, and this appendix will briefly cover the simulation methods used across all of the projects.

A.1 Geometrical Structures and Boundaries

The first step in any FEM simulation is to construct a model of the structure. A simple example is shown in Fig. A.1, in which the top and bottom features are set up as perfectly matched layers (PML), the function of which will be introduced in the next section. The center part is used to construct the filter, and can comprise an arbitrarily complicated geometric organization. The remaining two parts represent the top coating and the substrate. If there is no coating on the top of the filter (which indicates that the top is air), then semi-infinite conditions can be applied [2]. In this case, perfect electric conductor (PEC) and perfect magnetic conductor (PMC) conditions are applied to the layer or boundary perpendicular to the direction of electric field and magnetic field, respectively.

Material properties can be assigned to the various layers in a 2-D model and to domains in the corresponding 3-D models. In COMSOL, there are three ways to specify a material property as follows:

© The Editor(s) (if applicable) and The Author(s), under exclusive license to Springer
Nature Singapore Pte Ltd. 2021
X. He et al., *Multispectral Image Sensors Using Metasurfaces*, Progress in Optical Science and Photonics 17, https://doi.org/10.1007/978-981-16-7515-7

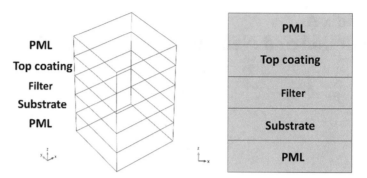

Fig. A.1 Example structures in 2D and 3D

- Interpolation Functions are used to describe properties defined by a table or file containing values of the function at discrete points, entered either manually or from a data file. The behavior between data points (and outside the range) is then specified.
- Analytic Functions are used to specify material properties as mathematical functions of one or more arguments, or as independent variables such as temperature or pressure, specified over a range.
- Piecewise Functions are useful for describing a material property that has different definitions on different intervals. The start and end points, function, and the interpolation and extrapolation methods have to be specified for each interval.

For example, refractive index is a very important material property in the design of optical filters, and many of the basic properties are assigned as part of the COMSOL materials library. However, the characteristics of a given material can vary widely when deposited with different methods and under varying conditions. Therefore, it will always be better (i.e., result in a more realistic simulation), to measure the exact material property using an appropriate method (e.g., ellipsometry), then import these data into the simulator. Further, although a single value is sometimes used for parameters such as refractive index (RI), Rakić's data [3] indicates that it is non-linearly dependant on the incident wavelength [4] and can vary widely over the visible and NIR ranges. Together, this indicates that the appropriate method to deal with RI is to import the experimental values and use a non-linear interpolation function to evaluate the intermediate points. Of course, if the imported data contains sufficient points, a linear approximation may be sufficient. If the real and imaginary parts for a typical material, Al, are defined as Al_Re and Al_Im, respectively, and a variable (e.g., lambda) is also defined to hold the wavelength, the complex refractive index (Al_Re + i Al_Im) can be expressed as (Al_Re lambda [1/nm]) and (Al_Im lambda[1/nm]), which instructs the software to interpolate between the experimental values at a 1 nm interval. This dependency can then be set up in the materials section of the simulation file, and then assigned to the appropriate layers in the 2-D models or to the domains in 3-D models.

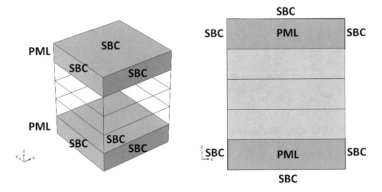

Fig. A.2 Example simulation model using SBC

A.1.1   Boundary Conditions

With COMSOL software, in contrast to other similar simulation systems (e.g. FDTD), several boundary conditions need to be set correctly. If this is not done, COMSOL may raise a warning flag, but might not and still give a result, albeit one that is incorrect. There are seven kinds of boundary conditions available in the RF module:

- perfectly matched layers (PML);
- scattering boundary conditions (SBC);
- port boundary conditions;
- perfect electric conductor (PEC) conditions;
- perfect magnetic conductor (PMC) conditions;
- periodic boundary conditions (PBC); and
- transition boundary condition (TBC).

These are described in the following subsections.

A.1.1.1   Perfectly Matched Layer (PML) and Scattering Boundary Condition (SBC)

A PML has the ability to absorb the light which may otherwise be reflected by a metal layer. For this reason, it is sometimes referred to as a "sponge" layer. The lack of such an absorbing layer would result in a secondary light source being transmitted back through the simulated structure, causes incorrect results. However, as mentioned before, this sort of incorrect model might be not be flagged by the software. To further absorb the reflected light, SBC should be applied around PMLs and PML should be applied at the top and bottom of the overall simulated blocks [5], as shown in Fig. A.2.

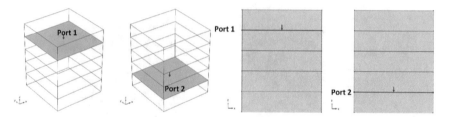

Fig. A.3 Example of using port boundary conditions

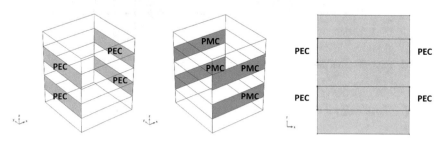

Fig. A.4 Example simulation model using PEC and PMC

A.1.1.2 Port Boundary Conditions

Port boundary conditions are used for the boundaries (in 2-D) or layers (3-D) to excite or stop light. A small number of predefined light sources are available: numeric, rectangular and periodic and users can also define their preferred source in the port mode setting. The direction of electric or magnetic field, propagation constant is given by $k = \frac{\omega}{c} = \frac{\frac{2\pi n c}{\lambda}}{c} = \frac{2\pi n}{c}$, where n is the refractive index of the excitation layer or domain and the phase direction can be defined. While applying PML at the top and bottom of a structure, the interior slit needs to be activated and the appropriate direction needs to be set as shown in Fig. A.3.

A.1.1.3 Perfect Electric Conductor (PEC) and Perfect Magnetic Conductor (PMC) Conditions

The theory of PEC conditions and PMC conditions can be described as follows [5]:

- For PMC: $n \times M = 0$, indicating that the electric field perpendicular to this layer or boundary will be absorbed.
- For PEC: $n \times E = 0$, so that the magnetic field perpendicular to this layer or boundary will be absorbed.

PEC and PMC can be used together for the semi-infinite layer/domain, assuming that the electric field of the incident light source is in the x direction, while the magnetic field is y direction. Therefore, PEC and PMC can be applied as shown in Fig. A.4 [2].

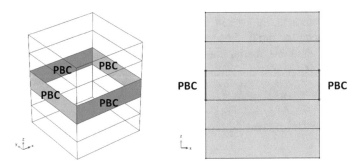

Fig. A.5 Example model using PBC

A.1.1.4   Periodic and Transition Boundary Conditions

PBC can be applied to the boundaries and layers that surround the filter structure, which repeats the block inside and can significantly reduce the computational time [5]. An example is shown in Fig. A.5. TBC can be used for the very layers within the structure that require a refractive index to be defined, but that are too thin to require meshing. Examples include the adhesion layer for the optical filter [5].

A.1.2   Meshing

Meshing plays an extremely important part in the operation of FEM simulation software such as COMSOL. It represents one of the main places that a user has control over the accuracy–performance trade off. A fine mesh will more accurately describe the underlying geometry of the material system, especially in the case of curved surfaces or materials with non-linear or discontinuous properties. On the other hand, a coarse mesh will be computationally simpler and will run to completion much faster. This trade off is something that a FEM user has to be aware of and to manage carefully.

In a 2-D simulation, a triangular partitioning is typical as this guarantees full coverage of the simulated area without grossly over-complicating the mesh. Triangles are also the only geometric shape guaranteed to be able to approximate a surface with no gaps or holes. The exception to the almost universal use of a triangular mesh is the PMLs, generally applied to boundaries, for which a sweep mesh in a rectangular shape is used.

In 3-D structures, the PBC layers are divided into triangular pieces, while the main structure (except PMLs) are divided into small tetrahedrons. Overall, this results in fewer elements in the mesh and therefore a solution will be found in a shorter time. The PMLs use a 3-D sweep technique together with distribution function to mesh the layers between PML and main structure.

As a final comment, the overall objective of the FEM simulation is typically to determine the absolute value of the transmission efficiency, which is calculated using abs(emw.S21)^2 in the RF model (the variable *emw* is called *ewfd* in the wave optics model of COMSOL).

A.2 Nanofabrication Methods

The range of topics relating to the processing of nano-scale materials and structures is vast, so this section will focus on just those materials and techniques used in the work demonstrated in this book.

A.2.1 Optical Materials

This section very briefly examines the types of materials typically used in optical systems such as Fabry-Pérot based filters, photonic bandgap filters, plasmonic gratings, nanohole arrays, metallic nanoparticle and metallic apertures.

For multilayer structures, including Fabry-Pérot based filters, photonic bandgap filter and dielectric multilayers, the choice of the material primarily depends on the target operating wavelength(s). For example, if the optical filter is intended to work in the visible wavelength (from 400 to 700 nm), then dielectric materials such as SiO_2, MgO, Ta_2O_5, TiO_2 and others that are potentially transparent in this wavelength range [6] are an appropriate choice. Among these, TiO_2 can be processed to exhibit the largest refractive index, up to 3.2 in some cases [7]. This is significantly different to SiO_2 (1.42), and is therefore commonly used for multilayer structures that require and abrupt change in refractive index. The range of techniques in common use to deposit these materials include E-beam evaporation, sputtering, PECVD, ALD and ion beam deposition (IBD). For all multilayer structures, surface roughness is the biggest challenge. If the deposition of any layer is insufficiently flat, the layers above it will be affected will add their own surface roughness to make the situation incrementally worse. Of the options listed above, ALD will result in the smoothest surface, as it deposits the material in atomically thick layers. However, this results in the slowest deposition rate among those listed. E-beam evaporator appears to be most common deposition facility used within academic research centers, as it has the broadest range of applications (e.g., lift off processing) that cannot be achieved with other deposition facilities. Moreover, an E-beam evaporator can control the deposition rate and roughness well, but the layers may not condense closely, which would significantly decrease the optical properties of the multilayer based filter. The solution is to use an Ion assist (normally Argon gas) while doing the deposition [6].

For plasmonic optical filters, the choice and deposition of the metal materials are the most important part. Au, Ag and Al are common materials used for plasmonic filters operating in the visible and NIR wavelength (from 400 to 1000 nm)

[8]. Other common metals used in electronic fabrication, such as Chromium (Cr), nickel (Ni), tungsten (W), and titanium (Ti) are less well suited here as they result in much lower transmission efficiency and broader spectral responses due to their lossy optical properties [9]. For example, Au will not support resonant wavelengths smaller than 550 nm, and therefore cannot be used for generating blue and green colours. Silver will quickly oxidize, certainly within a few weeks, thereby changing its optical properties. Moreover, Au and Ag have poor adhesion on most substrates, such as quartz (SiO_2) and sapphire (Al_2O_3) and thus require an additional seed layer (such as Cr, Ti) to avoid this problem. These additional materials may reduce the transmission efficiency and widen the spectral bandwidth [10]. There have been some filter structures proposed and built without using a seed layer between the Au or Ag and the dielectric substrate, but the size of these has been quite small and not practical for imaging applications [11]. In contrast, aluminium is well known as a CMOS compatible material that does not need a seed layer, and colour filters made of Al also exhibit a high transmission efficiency of more than 40% [12, 13].

A.2.2   Lithographic Techniques

There are numerous approaches to performing lithographic patterning on photo resist, including creating the pattern on polymethyl methacrylate (PMMA) with HV electron beams (e.g. EBL). In this case, the sample will require development with MIBK diluted with IPA. It must be noted that EBL will exhibit proximity effects. The electron beam can penetrate through the resist to form the pattern, but it will not be stopped by the resist if the dose used is large. In this case, the electron beam will penetrate the resist and will be reflected off the substrate (depending on the particular substrate material) and hit the resist again, thereby making the pattern edges less well defined. Further, some resist materials can be chemically altered by the impact energy of the electron beam and will required a different developer to remove this new chemical resist. If this chemical change is non-uniform across the surface, holes may develop in the resist leading to poor results. It is therefore critical that the e-beam dose, the developer concentration as well as the development time are all carefully optimized [14, 15].

Thermal scanning probe lithography (t-SPL) systems [16, 17], such as the Nanofrazor [18], can directly write the pattern onto a resist such as polyphthalamide (PPA). The material is directly evaporated by an ultra-sharp heated scanning probe tip and therefore does not require high voltages. On the other hand, the system operates at high temperature, around 900°C, which has the potential to damage the underlying surfaces. In practice, however, it has been found that the heat tends not to transfer through the PPA to the surface. While directly patterning the resist does avoid the need for a development step, this is replaced by the requirement to remove the PPA residue, normally achieved using a process like deep reactive-ion etching (DRIE). The reaction that occurs in the DRIE chamber is a chemical reaction in the presence of various single or mixed gas streams with plasma assist, in which the etching rate can be controlled by varying the power and gas flow rate. Focused ion beam (FIB) is also an important nanofabrication technique. FIB usually has two source guns, one used for imaging in the same way as SEM, while the other is used for milling the nanostructure in the sample with a beam of ions such as gallium [19, 20]. Although it has the advantage of being a mask-less fabrication technique as with EBL, some gallium residue may remain, causing contamination of the sample. One additional benefit of FIB is its ability to cut through the sample, allowing a cross section image to be produced.

A.2.3 Thin Film Deposition

Plasma-enhanced chemical vapour deposition (PECVD) is one of the best ways to deposit amorphous (a-) material, such as a-SiO_2 or a-Si_3N_4. The precursors mixed with a gas plasma result in a very conformal deposition, which is more uniform and faster than sputter deposition and e-beam evaporation [21]. Further, a heater is available in most PECVD models that can help the material condense onto the sample. Sputter deposition uses a RF/AC/DC source to physically etch the material from the target with gas assistance and direct it to the sample [22].

As mentioned above, the alternative of using an e-beam evaporator is common in academic laboratories. EBL uses HV to mill the target material and evaporate it onto the sample, and will also support nanofabrication processes that require a lift off step.

A.2.3.1 Materials Characterization

Scanning electron microscope (SEM) is used for scanning samples with HV electron beams. The atoms within the sample will react with the electron beam to produce a signal to image sub-micron features of the sample. This microscopy function is critical to be able to determine if the various steps of patterning, etching and lift off have worked as required. However, the samples must be imaged in a vacuum chamber to avoid contamination [23, 24] and to allow the e-beam to to be focused to the necessary resolution.

Atomic force microscopy (AFM) can also perform sub-micron microscopy imaging. The AFM operates in a similar way to t-SPL but has a read-out function only. The key advantages of AFM is that it can work in an ambient environment, does not require a vacuum, and there is no need to fix the sample in position. It also does not suffer from diffraction limits [25]. The deposition thickness of thin film under different environment are not the same. For example, when using sputtering to deposit metal, Argon gas is needed to strike the plasma, and therefore the deposition pressure will be different, resulting in a variable deposition rate. The only way to know the exact thickness is to then use AFM to measure the sample. AFM can also be used to measure the roughness of the material surface, which has significant influence on its optical properties.

A final useful technique to investigate the sample properties is ellipsometry, so-called because it measures the ellipse of polarization generated when a polarized light beam reflects obliquely from the surface of a sample. As light reflected by different thin films will exhibit different phase information, ellipsometry uses light at varying angles to the sample to investigate optical properties such as refractive index [26]. For example, a material such as TiO_2 will exhibit different values of refractive index when deposited with or without oxygen assist, and ellipsometry can be used to find this slight difference within its diffraction limit.

References

1. COMSOL Inc. Multiphysics cyclopedia: the finite element method. https://www.comsol.com/multiphysics/finite-element-method. Accessed June 2021
2. R. Rajasekharan, E. Balaur, A. Minovich, S. Collins, T.D. James, A. Djalalian-Assl, K. Ganesan, S. Tomljenovic-Hanic, S. Kandasamy, E. Skafidas, D.N. Neshev, P. Mulvaney, A. Roberts, S. Prawer, Filling schemes at submicron scale: development of submicron sized plasmonic colour filters. Sci. Rep. **4**(1), 6435 (2014)
3. A.D. Rakić, Algorithm for the determination of intrinsic optical constants of metal films: application to aluminum. Appl. Opt. **34**(22), 4755–4767 (1995)
4. Refractive Index Database: Optical constants of Al (Aluminium). https://refractiveindex.info/?shelf=main&book=Al&page=Rakic. Accessed July 2021
5. COMSOL Inc. RF module user's guide. https://doc.comsol.com/5.5/doc/com.comsol.help.rf/RFModuleUsersGuide.pdf. Accessed Apr 2020
6. L.V. Rodríguez de Marcos, J.I. Larruquert, J.A. Méndez, J.A. Aznárez, Self-consistent optical constants of SiO_2 and Ta_2O_5 films. Opt. Mater. Express **6**(11), 3622–3637 (2016)
7. C. Hu, J. Liu, J. Wang, Z. Gu, C. Li, Q. Li, Y. Li, S. Zhang, C. Bi, X. Fan, W. Zheng, New design for highly durable infrared-reflective coatings. Light Sci. Appl. **7**
8. M.W. Knight, N.S. King, L. Liu, H.O. Everitt, P. Nordlander, N.J. Halas, Aluminum for plasmonics. ACS Nano. **8**(1), 834–840 (2014)
9. C. Ji, K.-T. Lee, T. Xu, J. Zhou, H.J. Park, L.J. Guo, Engineering light at the nanoscale: structural color filters and broadband perfect absorbers. Adv. Opt. Mater. **5**(20), 1700368 (2017)
10. M. Todeschini, A.B. da Silva Fanta, F. Jensen, J.B. Wagner, A. Han, Influence of ti and cr adhesion layers on ultrathin au films. ACS Appl. Mater. Interf. **9**(42), 37374–37385 (2017)
11. Y. Liang, S. Zhang, X. Cao, L. Yanqing, X. Ting, Free-standing plasmonic metal-dielectric-metal bandpass filter with high transmission efficiency. Sci. Rep. **7**(1), 4357 (2017)
12. S. Yokogawa, S.P. Burgos, H.A. Atwater, Plasmonic color filters for CMOS image sensor applications. Nano Lett. **12**(8), 4349–4354 (2012)
13. S.P. Burgos, S. Yokogawa, H.A. Atwater, Color imaging via nearest neighbor hole coupling in plasmonic color filters integrated onto a complementary metal-oxide semiconductor image sensor. ACS Nano **7**(11), 10038–10047 (2013)
14. A.N. Broers, A.C.F. Hoole, J.M. Ryan, Electron beam lithography–resolution limits. Microelectron. Eng. **32**(1), 131–142 (1996)
15. K.W. Lee, S.M. Yoon, S.C. Lee, W. Lee, I.M. Kim, C.E. Lee, D.H. Kim, Secondary electron generation in electron-beam-irradiated solids: Resolution limits to nanolithography. J. Korean Phys. Soc. **55**(4), 1720–1723 (2009)
16. F. Holzner, M. Zientek, P. Paul, A. Knoll, C. Rawlings, Scanning probe nanolithography system and method, Feb 19, 2019. US Patent 10,209,630

17. H. Wolf, C. Rawlings, P. Mensch, J.L. Hedrick, D.J. Coady, U. Duerig, A.W. Knoll, Sub-20 nm silicon patterning and metal lift-off using thermal scanning probe lithography. J. Vacuum Sci. Technol. B **33**(2), 02B102 (2015)
18. Heidelberg Instruments Mikrotechnik GmbH. NanoFrazor Explore. https://heidelberg-instruments.com/product/nanofrazor-explore. Accessed Apr 2020
19. T.L. Burnett, R. Kelley, B. Winiarski, L. Contreras, M. Daly, A. Gholinia, M.G. Burke, P.J. Withers, Large volume serial section tomography by Xe Plasma FIB dual beam microscopy. Ultramicroscopy **161**, 119–129 (2016)
20. J. Orloff, L.W. Swanson, M. Utlaut, Fundamental limits to imaging resolution for focused ion beams. J. Vacuum Sci. Technol. B Microelectron. Nanometer Struct. Process. Measur. Phenomena **14**(6), 3759–3763 (1996)
21. H.F. Sterling and R.C.G. Swann. Chemical vapour deposition promoted by R.F. discharge. Solid-State Electron. **8**(8), 653–654 (1965)
22. K. Ishii, High-rate low kinetic energy gas-flow sputtering system. J. Vacuum Sci. Technol. A **7**(2), 256–258 (1989)
23. D. McMullan, Scanning electron microscopy 1928–1965. Scanning **17**(3), 175–185 (1995)
24. P. Hirsch, M. Kässens, M. Püttmann, L. Reimer, Contamination in a scanning electron microscope and the influence of specimen cooling. Scanning **16**(2), 101–110 (1994)
25. G.K. Bennig, Atomic force microscope and method for imaging surfaces with atomic resolution, Feb 9, 1988. US Patent 4,724,318
26. A. Rothen, The ellipsometer, an apparatus to measure thicknesses of thin surface films. Rev. Sci. Instr. **16**(2), 26–30 (1945)

Printed in the United States
by Baker & Taylor Publisher Services